Representation Theory of Finite Groups

MARTIN BURROW

COURANT INSTITUTE OF MATHEMATICAL SCIENCES
NEW YORK UNIVERSITY

DOVER PUBLICATIONS, INC.
New York

In memory of my mother

Published in Canada by General Publishing Company, Ltd., 30 Lesmill Road, Don Mills, Toronto, Ontario.
Published in the United Kingdom by Constable and Company, Ltd., 3 The Lanchesters, 162–164 Fulham Palace Road, London W6 9ER.

This Dover edition, first published in 1993, is an unabridged, slightly corrected republication of the second, corrected printing (1971) of the work first published by Academic Press, New York, 1965.

Manufactured in the United States of America
Dover Publications, Inc., 31 East 2nd Street, Mineola, N.Y. 11501

Library of Congress Cataloging-in-Publication Data

Burrow, Martin.
Representation theory of finite groups / Martin Burrow.
p. cm.
Includes bibliographcal references and index.
Reprint. Originally published: New York : Academic Press, 1965 (1971 printing).
ISBN 0-486-67487-8 (pbk.)
1. Representations of groups. 2. Finite groups. I. Title.
[QA176.B87 1993]
512′.2—dc20 92-39061
CIP

Preface

The old representation theory of finite groups by matrices over the complex field was largely the work of G. Frobenius, together with significant contributions by I. Schur. Many of the important results of Frobenius were again found independently by W. Burnside whose book, "Theory of Groups" (1911), is now a classic. Burnside's intriguing use of group characters to obtain results on abstract groups earned great attention for the theory.

The next decisive influence on the development of representation theory was E. Noether's shift of emphasis to the study of the representation module (1929). Her point of view has produced valuable gains in an algebraic direction.

A major extension of the subject in the last 25 years has been the study of modular representations initiated by R. Brauer. This theory too has provided significant applications to the theory of finite groups. For example, Thompson and Feit have recently aroused great interest by their use of the modular theory in the study of the solvability of groups of odd order [25].

Until recently much of the material on modular representations has been accessible only through research articles and the lectures of its principal developers. There was no systematic account of the modular theory until the publication (1962) of Curtis and Reiner's "Representation Theory of Finite Groups and Associative Algebras." A treatment of the modular theory which is due to R. Brauer is given in the final chapter of this book. The remaining chapters contain the standard material of representation theory which is here treated consistently from the point of view of the representation module. The rudiments of linear algebra and a knowledge of the elementary concepts of group theory are useful, if not entirely indispensable, prerequisites for reading this book. A graduate student so equipped who wishes to acquire some knowledge of representation theory should not find the work too difficult to master on his own. Also the book might prove useful as supplementary reading for a course in group theory or in the applications of representation theory to Physics.

The author wishes to express heartfelt thanks to Miss Kate Winter in particular and to Mr. J. Koppelman for their invaluable assistance in proofreading.

New York MARTIN BURROW
1965

Contents

Chapter III. The Representation Theory of Finite Groups

Chapter IV. Applications of the Theory of Characters

Chapter V. The Construction of Irreducible Representations

Chapter VI. **Modular Representations**

Appendix

CHAPTER I

Foundations

1. Introduction

Nowadays it is natural for us to think of a group abstractly as a set of elements $\{a, b, c, ...\}$, which is closed under an associative multiplication and which permits a solution, for x and y, of any equations: $ax = b$, and $ya = b$. On the other hand, we regard a group, which is given in some concrete way, as a realization of an abstract group. This point of view is an inversion of the historical development of group theory which won the abstract concept from particular modes of representation.

Group theory began with finite permutation groups. Any arrangement of n objects in a row is called a permutation of the objects. If we select some arrangement as standard, then any other arrangement can be regarded as achieved from it by an operation of replacements: each object in the standard being replaced by that object which takes its place in the new arrangement. Thus if 123 is standard and 312 is another arrangement, then the replacements are $1 \to 3$, $2 \to 1$, and $3 \to 2$. We write compactly for this operation

$$\begin{pmatrix} 1 & 2 & 3 \\ 3 & 1 & 2 \end{pmatrix}.$$

If the replacements of two operations are performed in succession, we get an arrangement which could be achieved directly by a third operation, called the *product* of the two operations. For example,

$$\begin{pmatrix} 1 & 2 & 3 \\ 2 & 3 & 1 \end{pmatrix}\begin{pmatrix} 1 & 2 & 3 \\ 2 & 1 & 3 \end{pmatrix} = \begin{pmatrix} 1 & 2 & 3 \\ 1 & 3 & 2 \end{pmatrix}.$$

Here we have proceeded from left to right. The product of operations is associative and any set of operations which form a group is a permutation group.

If we have n objects 1, 2, ..., n, then there are $n!$ arrangements and hence $n!$ operations are possible, including the identity:

$$\begin{pmatrix} 1 & 2 & \cdots & n \\ 1 & 2 & \cdots & n \end{pmatrix}.$$

They form a group S_n, the *symmetric group* on n symbols. Every permutation group on n symbols is a subgroup of S_n. In a remarkable application of a group theory in its infancy Galois showed that every algebraic equation possesses a certain permutation group on whose structure its properties depend.

Cayley discovered the abstract group concept. A theorem of his asserts that every abstract group with a finite number of elements can be realized as a group of permutations of its elements. Thus if $G = \{a, b, c, ..., g, ...\}$ is the abstract group, then the element of the permutation group P which corresponds to g is the set of replacements $a \to ag$, $b \to bg$, $c \to cg$, ..., or compactly:

$$g \sim \begin{pmatrix} x \\ xg \end{pmatrix}, \qquad x \text{ running over } G.$$

The groups G and P are isomorphic (see Appendix).

A generalization of the permutation group, and the next step historically, is the group of linear substitutions on a finite number of variables. In this case, if the variables are $x_1, ..., x_n$, the group operation consists of replacing each variable x_i by a linear combination of the variables; thus

$$(1.1) \qquad x_i \to x'_i = a_{1i}x_1 + a_{2i}x_2 + \cdots + a_{ni}x_n .$$

The coefficients a_{ij} are numbers, real or complex.

Substitutions are multiplied by carrying them out in succession. As an example let us find the product of the substitutions

$$(1.2) \qquad \begin{aligned} x' &= 2x - 3y \\ y' &= x - y \end{aligned} \qquad \text{and} \qquad \begin{aligned} x'' &= 3x' + 4y' \\ y'' &= 2x' + 3y'. \end{aligned}$$

Here we have written the first substitution as replacing unprimed variables by primed, and the second as replacing primed by double primed variables. Now in the second system substitute for x', y' from the first and get

$$\begin{aligned} x'' &= 3(2x - 3y) + 4(x - y) \\ y'' &= 2(2x - 3y) + 3(x - y) \end{aligned}$$

and so

$$\begin{aligned} x'' &= 10x - 13y \\ y'' &= 7x - 9y. \end{aligned} \tag{1.3}$$

This is the product substitution. Note that the use of primes is for distinction only. For instance (1.3) means that x is to be replaced by $10x - 13y$ and y by $7x - 9y$, irrespective of the single symbols, x'' and y'', which we use to denote these replacements. Going back to the general substitution (1.1) we see that it is entirely determined by the numbers a_{ij}, that is, by the matrix

$$\begin{pmatrix} a_{11} & a_{12} & \cdots & a_{1n} \\ \vdots & \vdots & \vdots & \vdots \\ a_{n1} & a_{n2} & \cdots & a_{nn} \end{pmatrix}. \tag{1.4}$$

The first row consists of the coefficients of x_1, taken in succession from the first, second, ..., nth equations, and in a similar way the ith row is formed from the coefficients of x_i. Note that (1.4) is the transpose of the array as it appears in (1.1). For example, in (1.2) we have the matrices

$$A = \begin{pmatrix} 2 & 1 \\ -3 & -1 \end{pmatrix}, \qquad B = \begin{pmatrix} 3 & 2 \\ 4 & 3 \end{pmatrix}.$$

Now

$$AB = \begin{pmatrix} 10 & 7 \\ -13 & -9 \end{pmatrix},$$

which is the matrix of the product substitution, so that in this case the correspondence between substitutions and their matrices is preserved by multiplication. It is easy to see that this is true in general. Thus, let X denote the row vector, or $1 \times n$ matrix, $(x_1, x_2, ..., x_n)$ and let X' be the same with primed variables. If (1.4) is denoted by A, then (1.1) can be written $X' = XA$. Now let $X'' = X'B$ be another substitution with matrix B. The product is the substitution $X'' = XAB$ and its matrix is AB.

If the substitutions form a group, the inverse substitution must exist and Eqs. (1.1) are solvable for the x_i in terms of the x_i'. This means that the determinant of the matrix (1.4) is not zero. Then A^{-1} exists and $X = X'A^{-1}$. We now see that a substitution group on a given set of variables is abstractly identical to a group of nonsingular matrices.

It is clear that substitutions, or matrices, admit a greater freedom of algebraic treatment than do permutations. For instance, matrices automatically generate a ring. Also, whereas permutation groups on a finite number of symbols are necessarily finite, substitutions allow us to deal with infinite groups. For example,

$$\begin{pmatrix} 1 & 0 \\ n & 1 \end{pmatrix}, \qquad n = 0, \pm 1, \pm 2, ...$$

is an infinite discrete group. Again, the substitutions

$$x' = x \cos \theta - y \sin \theta$$

$$y' = x \sin \theta + y \cos \theta$$

which leave the expression $x^2 + y^2$ invariant is an infinite continuous group. This is the *orthogonal group*. It is the group of rotations of the Cartesian plane about its origin.

Because of their algebraic flexibility it is natural to use matrices to represent abstract groups. Let us call a homomorphism of a group G into a group of $n \times n$ matrices a *representation of G of degree n*. This means that to each element g of G there corresponds a matrix $\sigma(g)$ and if x and y are any elements of G:

$$\sigma(xy) = \sigma(x)\sigma(y).$$

The representation is called *faithful* when the homomorphism σ is an isomorphism. When this is the case the correspondence is one-to-one and $\sigma(g) = I$, the identity matrix, if and only if $g = 1$, the identity of the group.

Frobenius proposed the question: Find all matrix representations of a finite group G. Let us make the following observations:

1. There always is a representation. For as we have seen there invariably is the permutation representation

$$g \sim \begin{pmatrix} x \\ xg \end{pmatrix}, \qquad x \text{ running through } G.$$

But, giving some fixed order $x_1, x_2, \ldots, x_n$ to the group elements, this merely expresses the linear substitutions

$$x_i \rightarrow x_{i'}$$

where $x_i g = x_{i'}$ and $i = 1, 2, \ldots, n$. Thus to each g corresponds a linear substitution and G is represented by a group of linear substitutions. Since $x_i g = x_i$ if and only if $g = 1$, the representation is faithful. The corresponding matrices give a faithful matrix representation of G. This is the *regular representation*. At the other extreme, so to speak, we have the one-representation i for which $i(g) = 1$, $\forall g \in G$.

2. Given any matrix representation we can find infinitely many. Thus if σ is a matrix representation and T is any fixed invertible matrix we can define $\tau(x) = T\sigma(x)T^{-1}$, $\forall x \in G$. Since

$$\begin{aligned} \tau(xy) = T\sigma(xy)T^{-1} = T\sigma(x)\sigma(y)T^{-1} &= T\sigma(x)T^{-1}T\sigma(y)T^{-1} \\ &= \tau(x)\tau(y), \end{aligned}$$

τ is a matrix representation of G. The new representation is brought about by a mere change of variable in the corresponding substitutions. To see this let the original variables be $X = (x_1, x_2, \ldots, x_n)$ and let new variables $Y = (y_1, y_2, \ldots, y_n)$ be related by

$$x_i = \sum_{j=1}^{n} t_{ji} y_j, \quad \text{or in matrices:} \qquad X = YT.$$

Then any substitution $X' = XA$, with matrix A, in the old variables becomes when expressed in the new: $Y'T = YTA$, and hence $Y' = YTAT^{-1}$, with matrix TAT^{-1}. Representations related as are σ and τ are said to be *equivalent* and are regarded as essentially the same representation. All representations equivalent to σ are clearly equivalent to each other and form an infinite class. Since they are all the same representation the task proposed by Frobenius can be narrowed to the survey of all inequivalent representations. We will write $\sigma \sim \tau$ or $\sigma \nsim \tau$ according as the two representations are or are not equivalent.

3. If σ and τ are two matrix representations of G of degrees s and t, respectively, and if $g \in G$, the matrix

$$\mu(g) = \begin{pmatrix} \sigma(g) & 0 \\ 0 & \tau(g) \end{pmatrix}$$

of degree $s + t$ is the direct sum of the matrices $\sigma(g)$ and $\tau(g)$. The 0 in the first row and that in the second represent, respectively, an $s \times t$ and a $t \times s$ matrix of zeros. Now

$$\mu(x)\mu(y) = \begin{pmatrix} \sigma(x) & 0 \\ 0 & \tau(x) \end{pmatrix}\begin{pmatrix} \sigma(y) & 0 \\ 0 & \tau(y) \end{pmatrix} = \begin{pmatrix} \sigma(x)\sigma(y) & 0 \\ 0 & \tau(x)\tau(y) \end{pmatrix}$$

$$= \begin{pmatrix} \sigma(xy) & 0 \\ 0 & \tau(xy) \end{pmatrix} = \mu(xy).$$

Hence μ is a representation. It is called the *direct sum* of σ and τ and we write $\mu = \sigma \oplus \tau$. Thus, given two representations, possibly the same, we can find a third by adding them directly. Conversely, let μ be a representation of degree $s + t$ and $\forall x \in G$ suppose that the matrix $\mu(x)$ has the form

$$\begin{pmatrix} A(x) & 0 \\ 0 & B(x) \end{pmatrix}$$

where $A(x)$ and $B(x)$ are, respectively, $s \times s$ and $t \times t$ matrices depending on the element x. As before, the 0's are $s \times t$ and

$t \times s$ zero matrices. We now define σ and τ by: $\sigma(x) = A(x)$, $\tau(x) = B(x)$, $\forall x \in G$. Since $\mu(xy) = \mu(x)\mu(y)$,

$$\begin{pmatrix} A(xy) & 0 \\ 0 & B(xy) \end{pmatrix} = \begin{pmatrix} A(x) & 0 \\ 0 & B(x) \end{pmatrix}\begin{pmatrix} A(y) & 0 \\ 0 & B(y) \end{pmatrix}$$
$$= \begin{pmatrix} A(x)A(y) & 0 \\ 0 & B(x)B(y) \end{pmatrix}.$$

Therefore $\sigma(xy) = \sigma(x)\sigma(y)$ and $\tau(xy) = \tau(x)\tau(y)$ so that σ and τ are representations. The representation μ is said to be *decomposable* and σ and τ are its *components*. If $\nu \sim \mu$, ν is also said to be decomposable. A representation which is not decomposable is *indecomposable*.

The basic question can now be more sharply put: Find all inequivalent indecomposable representations of a finite group G. Are there only a finite number of different ones? If a component of a decomposable representation is itself decomposable we get a further decomposition. Continuing in this way we can decompose any given representation into a finite number of indecomposable components. Are these unique, or can a different set be obtained by carrying out the decomposition differently? In due course we shall see these natural questions answered completely by the theory.

Historically, Frobenius started representation theory with the group determinant. Though it will play no role in our treatment we mention it now. If $A_1 = I, A_2, \ldots, A_n$ are the elements of a finite matrix group in any fixed order and $x_1, x_2, \ldots, x_n$ are an equal number of indeterminate variables, we can form the group matrix

$$M = x_1 A_1 + x_2 A_2 + \cdots + x_n A_n .$$

Its coefficients consists of linear combinations of the x_i; for example, at the intersection of the ith row and jth column we have

$$m_{ij} = \sum_{k=1}^{n} a_{ij}^{k} x_k$$

where a_{ij}^k is the coefficient in the same position in the matrix A_k. The determinant of this matrix is the *group determinant*. It is a polynomial of the same degree as the representation. It can be shown that this polynomial is factorable if and only if the group is decomposable and that then its irreducible factors are the group determinants of the indecomposable components. Moreover, the same transformation T that decomposes the group will transform M into the corresponding group matrices (see the Exercises).

4. The property of being indecomposable depends on the field of the representation. For example, the matrix group

$$\left\{\begin{pmatrix}1 & 0\\0 & 1\end{pmatrix}, \begin{pmatrix}-1 & 1\\-1 & 0\end{pmatrix}, \begin{pmatrix}0 & -1\\1 & -1\end{pmatrix}\right\}$$

is indecomposable over the rational field: there is no matrix T with rational coefficients so that

$$T\begin{pmatrix}-1 & 1\\-1 & 0\end{pmatrix}T^{-1} = \begin{pmatrix}\alpha & 0\\0 & \beta\end{pmatrix}.$$

This is easily seen by equating for the trace and determinant on both sides; since transforming a matrix as on the left of the equation does not alter these quantities, we have $\alpha + \beta = -1$, $\alpha\beta = 1$, leading to the quadratic equation $\alpha^2 + \alpha + 1 = 0$, whose roots are cube roots of unity. On the other hand, if complex values are permitted,

$$\begin{pmatrix}\omega & 1\\\omega^2 & 1\end{pmatrix}\begin{pmatrix}-1 & 1\\-1 & 0\end{pmatrix}\begin{pmatrix}\omega & 1\\\omega^2 & 1\end{pmatrix}^{-1} = \begin{pmatrix}\omega & 0\\0 & \omega^2\end{pmatrix}$$

where ω is the cube root of unity $\omega = (-1 + \sqrt{3}i)/2$. It is interesting to note that in this example the group determinant is $x^2 + y^2 + z^2 - xy - xz - yz$ which is irreducible in the rational or real field but can be factored thus:

$$(x + \omega y + \omega^2 z)(x + \omega^2 y + \omega z)$$

in the complex field.

Again consider the group

$$G = \left\{ \begin{pmatrix} 1 & 0 \\ 0 & 1 \end{pmatrix}, \begin{pmatrix} 1 & 0 \\ 1 & 1 \end{pmatrix} \right\}$$

with its coefficients in the prime field of characteristic 2, that is, in the field of two elements 0, 1 where addition and multiplication is performed modulo 2. If

$$T \begin{pmatrix} 1 & 0 \\ 1 & 1 \end{pmatrix} T^{-1} = \begin{pmatrix} a & 0 \\ 0 & b \end{pmatrix},$$

then on equating the trace and determinant on both sides we should have $a + b = 0$ $ab = 1$, yielding $a = b = 1$. Thus the right side is the identity matrix. This is impossible and so G is indecomposable. Moreover, unlike the previous example, G stays indecomposable even if the field be extended. Such a representation which remains indecomposable in any extension of the field is called *absolutely indecomposable*. If at the outset we consider the field of representation to be algebraically closed (and so incapable of further algebraic extension) or at any rate sufficiently large, then we may confine our attention to absolutely indecomposable representations. In general we will adopt this simplification. It turns out that the representation theory in fields of characteristic p is very different from the case at characteristic zero, if the order of the group is divisible by p.

5. There is a weaker concept than decomposability. Thus let μ be a matrix representation which for every $x \in G$ has the form

$$\mu(x) = \begin{pmatrix} A(x) & 0 \\ I(x) & B(x) \end{pmatrix} \tag{1.5}$$

in which $A(x)$, $B(x)$ are, respectively, $s \times s$ and $t \times t$ matrices depending on x. $I(x)$ is a $t \times s$ matrix depending on x and 0 is the $s \times t$ zero matrix. Defining $\sigma(x) = A(x)$ and $\tau(x) = B(x)$ it is easy to see as in Observation 3 that σ and τ are representations of G. The representation μ, and any equivalent representation, is said to be *reducible*. A representation that is not reducible is *irreducible*. The representations σ and τ are *constituents* of μ. If either is irreducible it is an *irreducible constituent*. A representation which is decomposable is clearly reducible [$I(x) = 0$ in (1.5)]. On the other hand, the second example of observation 4 shows that a representation which is reducible need not be decomposable. If the characteristic of the field is zero, or is not a divisor of the order of the group, this cannot happen. We shall see that in these cases every reducible representation is decomposable provided the field is taken large enough, and then the constituents are components as well.

If σ and τ are themselves reducible, they yield constituents of smaller degree. Continuing in this way we obtain a set of irreducible representations $\sigma_1, \sigma_2, \ldots, \sigma_k$ as constituents of μ. It will be shown that these are uniquely determined up to equivalence for each representation μ. Moreover, we shall see that the set of distinct irreducible representations of a finite group in a given field is an invariant of the group. They are finite in number and any irreducible constituent of a representation must be one of them. They have been called the building blocks for the representations of the group.

If the indecomposable components of a representation are also irreducible the representation is called *completely reducible.*

Example. The symmetric group S_3 of order 6 is generated by all products of the permutations

$$a = \begin{pmatrix} 1 & 2 & 3 \\ 2 & 1 & 3 \end{pmatrix} \quad \text{and} \quad b = \begin{pmatrix} 1 & 2 & 3 \\ 3 & 1 & 2 \end{pmatrix}.$$

Expressed as substitutions this gives

$$\begin{aligned} x_1' &= x_2 & \qquad & & x_1' &= x_3 \\ x_2' &= x_1 & & \text{and} & x_2' &= x_1 \\ x_3' &= x_3 & & & x_3' &= x_2 \end{aligned}$$

so that the corresponding matrix representation is

$$\nu(a) = \begin{pmatrix} 0 & 1 & 0 \\ 1 & 0 & 0 \\ 0 & 0 & 1 \end{pmatrix}, \qquad \nu(b) = \begin{pmatrix} 0 & 1 & 0 \\ 0 & 0 & 1 \\ 1 & 0 & 0 \end{pmatrix}.$$

Now ν is reducible, for if

$$T = \begin{pmatrix} 1 & 1 & 1 \\ 1 & -1 & 0 \\ 1 & 0 & -1 \end{pmatrix} \quad \text{and} \quad \mu(g) = T\nu(g)T^{-1}, \quad \forall g \in S_3,$$

then

$$\mu(a) = \begin{pmatrix} 1 & 0 & 0 \\ 0 & -1 & 0 \\ 0 & -1 & 1 \end{pmatrix}, \quad \mu(b) = \begin{pmatrix} 1 & 0 & 0 \\ 0 & -1 & 1 \\ 0 & -1 & 0 \end{pmatrix}.$$

Thus $\nu \sim \sigma \oplus \tau$, where

$$\sigma(a) = \sigma(b) = 1, \quad \tau(a) = \begin{pmatrix} -1 & 0 \\ -1 & 1 \end{pmatrix}, \quad \text{and} \quad \tau(b) = \begin{pmatrix} -1 & 1 \\ -1 & 0 \end{pmatrix}.$$

Now τ is irreducible, for if there were a matrix L such that

$$(*) \qquad \begin{aligned} L\begin{pmatrix} -1 & 0 \\ -1 & 1 \end{pmatrix}L^{-1} &= \begin{pmatrix} x & 0 \\ * & y \end{pmatrix} \\ L\begin{pmatrix} -1 & 1 \\ -1 & 0 \end{pmatrix}L^{-1} &= \begin{pmatrix} u & 0 \\ * & v \end{pmatrix} \end{aligned}$$

then, by equating traces and determinants, we would get $x = \pm 1$, $y = \mp 1$, $u = \omega$ or ω^2, $v = \omega^2$ or ω. Here ω is a cube root of unity. But on adding Eqs. $(*)$ we have

$$L\begin{pmatrix} -2 & 1 \\ -2 & 1 \end{pmatrix}L^{-1} = \begin{pmatrix} \pm 1 + \alpha & 0 \\ * & \mp 1 + \beta \end{pmatrix}, \qquad \begin{aligned} \alpha &= \omega \quad \text{or} \quad \omega^2 \\ \beta &= \omega^2 \quad \text{or} \quad \omega. \end{aligned}$$

This is impossible since the determinant of the matrix on the left is zero and that of the matrix on the right is not. Thus there are no relations $(*)$ and τ is irreducible. Clearly σ is irreducible and so ν is completely reducible.

The representation τ lies in the rational field. We have seen that it remains irreducible even when transformed by a matrix L which is allowed to have complex coefficients. A representation which, like τ, remains irreducible in any extension of the field of representation is called *absolutely irreducible*. The remark of Observation 4 concerning absolutely indecomposable representations applies here as well.

The set $L_n(F)$ of all nonsingular $n \times n$ matrices with coefficients in a field F form a multiplicative group called a *linear group*. We conclude this section by restating formally:

(1.6) **Definition.** *A matrix representation of degree n of a group G is a homomorphism σ of G into the linear group $L_n(F)$. This means that to each $x \in G$ there corresponds an $n \times n$ matrix $\sigma(x)$ with entries in F and, for all $x, y \in G$,*

$$\sigma(xy) = \sigma(x)\sigma(y).$$

The representation is faithful if σ is an isomorphism.

2. Group Characters

The matrices representing a group, besides the readiness with which they admit calculation, allow the introduction of numerical functions on the group. These functions, called characters, play an important part in the theory. We recall that if $A = (a_{ij})$, $i, j = 1, 2, \ldots, n$, is any matrix, then tr A, the trace of A, is

$$\operatorname{tr} A = \sum_{i=1}^{n} a_{ii} \, .$$

It can be seen by direct calculation that if T is any other matrix,

$$\operatorname{tr} AT = \operatorname{tr} TA$$

and then, if T is nonsingular so that T^{-1} exists,

$$\operatorname{tr} T^{-1}AT = \operatorname{tr} A.$$

(2.1) **Definition.** *The character of the representation* σ *of a group* G *is the function* χ^σ *on* G *to the field of representation* F *given by*

$$\chi^\sigma(g) = \text{tr } \sigma(g), \qquad \forall g \in G.$$

If $\mu \sim \sigma$, then by definition, $\forall g$ we have $\mu(g) = T\sigma(g)T^{-1}$ for some fixed matrix T. But then

$$\chi^\mu(g) = \text{tr } \mu(g) = \text{tr } T\sigma(g)T^{-1} = \text{tr } \sigma(g) = \chi^\sigma(g).$$

This shows that equivalent representations have the same character. Now if μ is a reducible representation, for some fixed matrix T

$$T\mu(x)T^{-1} = \begin{pmatrix} \sigma(x) & 0 \\ J(x) & \tau(x) \end{pmatrix}, \qquad \forall x \in G.$$

Then taking the trace on both sides we get

$$\chi^\mu(x) = \chi^\sigma(x) + \chi^\tau(x), \qquad \forall x \in G,$$

showing that the character of a reducible [decomposable, if $J(x) \equiv 0$] representation is the sum of the characters of its constituents (components). Thus every character is the sum of irreducible characters, that is, characters of irreducible representations.

A group can be partitioned into equivalence classes of conjugate elements, where g and h are in the same class if and only if there is an element t such that $g = t^{-1}ht$. Since

$$\begin{aligned} \chi^\sigma(g) = \chi^\sigma(t^{-1}ht) &= \text{tr } \sigma(t^{-1}ht) = \text{tr}(\sigma(t))^{-1}\sigma(h)\sigma(t) \\ &= \text{tr } \sigma(h) = \chi^\sigma(h), \end{aligned}$$

we see that the character is a class function. It has the same value for elements of the same class.

3. Representation Modules

Representations can be given a more algebraic formulation and thus fitted into a wider context by means of the concept of a representation module. This point of view is due to E. Noether. At the same time we will consider the more general question of the representations of a ring. First we introduce the double module.

(3.1) **Definition.** *Let F be a commutative field and R a ring with an identity. A set M of elements is an* ***F-R module*** *if:*

1. M is an additive abelian group.
2. $\forall f \in F$, $\forall m \in M$, $\forall r \in R$, there are unique elements fm and $mr \in M$. Moreover $(fm)r = f(mr)$.
3. $f(m + m') = fm + fm'$.
4. $(f + f')m = fm + f'm$.
5. $(ff')m = f(f'm)$.
6. $(m + m')r = mr + m'r$.
7. $m(r + r') = mr + mr'$.
8. $m(rr') = (mr)r'$.
9. $1m = m = m\mathbf{1}$, 1 and $\mathbf{1}$ being the identities of F and R.

In the language of group theory M is an abelian group with two distinct sets of operators. If we disregard the operators R, then M is an *F-module* or a *vector space over F*. When R is an algebra over F, so that $fr = rf \in R$ is defined for $f \in F$ and $r \in R$, there is the added condition:

10. $m(fr) = (fm)r$.

Finally, throughout most of our discussion we will require:

11. M has a finite F-basis.

Since M is a finite-dimensional vector space over F there is a close connection between linear transformations in M and matrices. Let us recall:

(3.2) **Definition.** *An F-endomorphism of an F-module M is a mapping σ of M into itself which assigns to each $m \in M$ a unique element $m\sigma \in M$ and which satisfies the rules*

$$(fm)\sigma = f(m\sigma) \tag{3.3}$$

$$(m + m')\sigma = m\sigma + m'\sigma. \tag{3.4}$$

If σ and τ are two endomorphisms their sum and product are defined by

(3.5) $\quad m(\sigma + \tau) = m\sigma + m\tau \qquad \text{and} \qquad m(\sigma\tau) = (m\sigma)\tau.$

The set of all endomorphisms of M form a ring $\mathfrak{L}$. The ring $\mathfrak{L}$ of F-endomorphisms of M is actually an algebra over F for we can define $f\sigma = \sigma f$ as the mapping of M into M given by $m(f\sigma) = (fm)\sigma$. Then $f\sigma \in \mathfrak{L}$. $\mathfrak{L}$ is the ring of all linear transformations of the vector space M over F. If M is of finite dimension n over F, $\mathfrak{L}$ can be represented analytically by the ring $\mathfrak{M}_n$ of all $n \times n$ matrices over F in the following way. Let $B = \{m_1, m_2, ..., m_n\}$ be an ordered basis of M over F. Then

(3.6) $$m_i\sigma = \sum_{j=1}^{n} f_{ij}^{\sigma} m_j, \qquad i = 1, 2, ..., n,$$

and the n^2 elements $f_{ij}^{\sigma} \in F$. Thus σ is associated with the $n \times n$ matrix $S = (f_{ij}^{\sigma})$. Conversely, if S is given, a mapping σ of the basis can be defined by (3.6) and extended to M by using (3.3) and (3.4) as definitions. Thus there is a one-to-one correspondence between linear transformations σ and $n \times n$ matrices S. If $\sigma \leftrightarrow S$ and $\tau \leftrightarrow T$, it is easily checked that $\sigma + \tau \leftrightarrow S + T$ and $\sigma\tau \leftrightarrow ST$, so that $\mathfrak{L} \cong \mathfrak{M}_n$.

(3.7) **Definition.** *A representation of a ring R is a ring homomorphism ρ of R into $\mathfrak{L}$, the ring of linear transformations of an F-module M.*

This means that to each $r \in R$ there corresponds a unique element $\rho(r) \in \mathfrak{L}$ such that

(3.8) $$\rho(r + r') = \rho(r) + \rho(r')$$

(3.9) $$\rho(rr') = \rho(r)\rho(r').$$

On taking a basis for M, $\rho(r)$ is associated with a matrix $\check{M}(r)$ and these relations hold for the corresponding matrices.

An F-R module may be called a *representation module* or, *representation space*, of R, for we have:

(3.10) **Lemma.** *To each F-R module M there is a unique representation of the ring R and conversely.*

Proof. Let $r \in R$. Define the mapping $\mu(r) : M \to M$ as follows:

$$m\mu(r) = mr, \qquad \forall m \in M.$$

Then

$$(fm)\mu(r) = (fm)r = f(mr) = f(m\mu(r))$$

and

$$(m + m')\mu(r) = (m + m')r = mr + m'r = m\mu(r) + m'\mu(r).$$

Thus $\mu(r)$ satisfies (3.3) and (3.4) and is therefore an F-endomorphism of M as an F-module. Hence $\mu(r) \in \mathfrak{L}$. Moreover, $\forall m \in M$,

$$\begin{aligned} m\mu(r + r') &= m(r + r') = mr + mr' = m\mu(r) + m\mu(r') \\ &= m(\mu(r) + \mu(r')), \qquad \text{the final step by (3.5),} \end{aligned}$$

and

$$\begin{aligned} m\mu(rr') &= m(rr') = (mr)r' = (m\mu(r))\mu(r') \\ &= m(\mu(r)\mu(r')), \qquad \text{the final step by (3.5).} \end{aligned}$$

This shows that

$$\mu(r + r') = \mu(r) + \mu(r')$$

$$\mu(rr') = \mu(r)\mu(r')$$

and hence that μ is a representation of R. Conversely, if $\forall r \in R$, $\mu(r)$ is an F-endomorphism of an F-module M and satisfies the last two equations, we can turn M into an F-R module by defining: $mr = m\mu(r)$. The proof is now complete.

This result enables us to turn our attention from the representation to the corresponding representation module. Since this is an abelian group with operators, certain fundamental results from group theory can be applied. This is done in the next section.

4. Application of Ideas and Results from Group Theory

Since a representation module is an abelian group with operators it is natural to investigate the meaning, for representations, of such group theoretical notions as: admissable subgroup, factor group, direct product (sum). First we have the natural:

(4.1) **Definition.** *Two representations μ and ν of a ring R are equivalent (or similar) if their* ***corresponding representation modules M and N are operator isomorphic****; that is, if there is an isomorphism $\tau : M \to N$ such that $\forall m \in M$ there exists a unique $m\tau \in N$ and*

$$(m + m')\tau = m\tau + m'\tau$$

$$(fmr)\tau = f(m\tau)r, \qquad \forall f \in F, \quad \forall r \in R.$$

We are going to show that this definition of the equivalence of representations is in accord with the one introduced earlier. Now $\forall m \in M$

$$m(\mu(r)\tau) = (m\mu(r))\tau = (mr)\tau = (m\tau)r = (m\tau)\nu(r) = m(\tau\nu(r))$$

and hence

$$\text{(4.2)} \qquad \mu(r)\tau = \tau\nu(r), \qquad \forall r \in R.$$

Because of (4.2) τ is said to intertwine the representations μ and ν and is called an *intertwining mapping*. Since τ is one-to-one onto, τ^{-1} exists and $\mu(r) = \tau\nu(r)\tau^{-1}$.

Now let $B = \{m_1, m_2, \ldots, m_k\}$ and $B' = \{n_1, n_2, \ldots, n_k\}$ be ordered bases of M and N. Then

$$(*) \qquad \begin{aligned} m_i\mu(r) &= m_i r = \sum_{j=1}^{k} f_{ij}^{\mu}(r) m_j, \qquad f_{ij}^{\mu}(r) \in F, \\ n_i\nu(r) &= n_i r = \sum_{j=1}^{k} f_{ij}^{\nu}(r) n_j, \qquad f_{ij}^{\nu}(r) \in F, \end{aligned}$$

and in each case $i = 1, 2, ..., k$. Thus we have the correspondences

$$\mu(r) \leftrightarrow \check{M}(r) = (f_{ij}^{\mu}(r)) \qquad \text{and} \qquad \nu(r) \leftrightarrow \check{N}(r) = (f_{ij}^{\nu}(r))$$

where $\check{M}(r)$ and $\check{N}(r)$ are each $k \times k$ matrices depending on r. If now $m_i\tau = \Sigma_{j=1}^{k} t_{ij}n_j$, $i = 1, 2, ..., k$, then τ is associated with a matrix $T = (t_{ij})$. Applying τ to the first equation of $(*)$ and using (4.2):

$$m_i\mu(r)\tau = m_i\tau\nu(r) = \sum_{j=1}^{k} t_{ij}n_j\nu(r) = \sum_{j=1}^{k}\sum_{l=1}^{k} t_{ij}f_{jl}^{\nu}(r)n_l$$

$$= \sum_{j=1}^{k} f_{ij}^{\mu}(r)m_j\tau = \sum_{j=1}^{k}\sum_{l=1}^{k} f_{ij}^{\mu}(r)t_{jl}n_l\,.$$

Comparing the fourth term with the last term we see that the entry at the intersection of the ith row and lth column of $T\check{N}(r)$ is the same as the corresponding entry of $\check{M}(r)T$. This is true $\forall i, l$ and so

$$\check{M}(r)T = T\check{N}(r),$$

or

$$\check{M}(r) = TN(r)T^{-1}, \qquad \forall r \in R,$$

yielding our former definition of the equivalence of matrix representations.

Let us recall that S is an *admissable submodule* of an F-R module M if S is a subgroup of the group M and if $\forall f \in F$, $\forall s \in S$, $\forall r \in R$, $fsr \in S$. Then S is itself an F-R module and provides a representation. Moreover, since M is abelian, S is a normal subgroup. The factor group M/S is also an F-R module if we define the action of the operators thus: $f\bar{m}r = \overline{fmr}$, where $\bar{m}$ denotes the class of m modulo S. How are the representations provided by S and M/S related to that provided by M? We can find a basis $\{m_1, m_2, ..., m_a, m'_1, m'_2, ..., m'_b\}$ of M such that the first a terms are a basis of S. Such a basis is said to be *adapted* (or accommodated) to S. With this basis we find the corresponding matrix representation:

$$
\begin{aligned}
m_1 r &= f_{11} m_1 + \cdots + f_{1a} m_a \\
&\ \vdots \\
m_a r &= f_{a1} m_1 + \cdots + f_{aa} m_a \\
m_1' r &= \check{f}_{11} m_1 + \cdots + \check{f}_{1a} m_a + f_{11}' m_1' + \cdots + f_{1b}' m_b \\
&\ \vdots \\
m_b' r &= \check{f}_{b1} m_1 + \cdots + \check{f}_{ba} m_a + f_{b1}' m_1' + \cdots + \check{f}_{bb}' m_b \,.
\end{aligned}
\tag{4.3}
$$

This gives the matrix representation

$$
\mu(r) = \begin{pmatrix} \mu_1(r) & 0 \\ J(r) & \mu_2(r) \end{pmatrix}
\tag{4.4}
$$

in which $\mu_1(r) = (f_{ij})$, $i, j = 1, 2, \ldots, a$, and $\mu_2(r) = (f'_{ij})$, $i, j = 1, 2, \ldots, b$. μ_2 is the representation corresponding to the representation module M/S. This can be seen by taking the last b equations of (4.3) modulo S.

If M is the direct sum of two admissable submodules, $M = S \oplus T$, then a basis adapted to this decomposition will yield a matrix (4.4) having $J(r) \equiv 0$. Here μ_1 and μ_2 are representations arising from the modules S and T, respectively.

We can now state formally:

(4.5) **Definition.** (a) *A representation μ of a ring R and its corresponding module M are* ***reducible*** *if M possesses a proper ($\neq 0$, M) admissable submodule S. The representation μ_1 corresponding to S is a* ***top constituent*** *of μ. The representation μ_2 corresponding to M/S is a* ***bottom constituent*** *of μ. If M has no proper admissable submodule it is irreducible and μ is an* ***irreducible representation****.*

(b) *A representation μ of a ring R and its corresponding module M are* ***decomposable*** *if M is a direct sum $M = S \oplus T$, and S, T are proper F-R submodules. The modules S, T and their correspon-*

ding representations μ_1 and μ_2 are called **components** *of M and μ, respectively.* **Indecomposable** *means not decomposable.*

(c) *A representation μ of a ring R and its corresponding module M is* **completely reducible** *if M is a direct sum $M = M_1 \oplus \cdots \oplus M_k$ and the components M_i are irreducible.*

A group G with operators is said to satisfy the *descending chain condition* if every sequence

$$G_1 \supset G_2 \supset G_3 \supset \cdots \supset \cdots$$

of admissable subgroups G_i contains only a finite number of distinct terms. Thus, if the inclusions are proper, such a sequence must terminate. G satisfies the *ascending chain condition* if every sequence

$$G_1 \subset G_2 \subset G_3 \subset \cdots$$

of admissable subgroups G_i contains only a finite number of distinct terms. Again, if the inclusions are proper, the chain must end with G at most.

Since our representation modules M, and their admissable submodules, are finite-dimensional vector spaces over F, any submodule must have a smaller dimension than that of another in which it is properly contained. Hence *both chain conditions are satisfied by the modules M.*

It is not hard to show that the chain conditions are equivalent, respectively, to the *maximum (minimum) condition*: every nonempty set of admissable subgroups has a maximal (minimal) member; that is, a member which is contained in (contains) no other member of the set.

Now let M be a reducible F-R module. Then there is a proper F-R submodule M'' such that

$$M \supset M'' \supset 0.$$

If M'' is itself reducible there is a proper submodule M''' such that $M'' \supset M''' \supset 0$. If M/M'' is reducible it contains a submodule K: $M/M'' \supset K$. This implies the existence of a submodule M' such that $M \supset M' \supset M''$, and $K \cong M'/M''$.

Then the original series can be extended to

$$M \supset M' \supset M'' \supset M''' \supset 0.$$

In the same way we may insert a submodule between any two consecutive modules in this series as long as the factor module of the one by the other is reducible. Repeated insertions will finally give a series

$$0 = M_0 \subset M_1 \subset M_2 \subset \cdots \subset M_k = M \tag{4.6}$$

which is incapable of further refinement. It is called a *composition series of length k*. The M_i/M_{i-1} are *factors* of the series. Let some other reduction of M give a second composition series:

$$0 = M_0' \subset M_1' \subset M_2' \subset \cdots \subset M_s' = M. \tag{4.7}$$

The Jordan-Hölder theorem for groups with operators for which the chain conditions are valid states that $k = s$ and that the factors M_i/M_{i-1} of (4.6) are operator isomorphic to the factors M_j'/M_{j-1}' of (4.7) in some order. Since the irreducible factors M_i/M_{i-1} are the representation modules for the irreducible constituents $\mu_1, \mu_2, \ldots, \mu_k$ of the representation μ which correspondes to the module M we have:

(4.8) **Theorem.** *Any representation μ of a ring R has a fixed number k of irreducible constituents $\mu_1, \mu_2, \ldots, \mu_k$. They are unique up to equivalence and order of arrangement.*

The case is similar if M is decomposable. Then $M = M_1 \oplus M_2$ and repeated decomposition of decomposable summands will give a Remak decomposition

$$M = M_1 \oplus M_2 \oplus \cdots \oplus M_k \tag{4.9}$$

in which each M_i is indecomposable. Let

$$M = M_1' \oplus M_2' \oplus \cdots \oplus M_s', \qquad M_j' \text{ indecomposable}, \tag{4.10}$$

be any other decomposition. The *Remak-Krull-Schmidt theorem* for groups with operators for which the chain conditions are valid asserts that $k = s$ and that the M_i are operator isomorphic to the M_j' in some order.

Since the M_i are representation modules for the indecomposable components μ_i of the representation μ, this result gives:

(4.11) **Theorem.** *Each representation μ of a ring R has a fixed number s of indecomposable components $\mu_1, \mu_2, \ldots, \mu_s$. These are unique up to equivalence and order of arrangement.*

It is instructive to look at a matrix representation of R which exhibits the irreducible constituents. Let B be an ordered basis of the representation module M accommodated to the series (4.6):

$$B = \{\overset{1}{m}_1, \ldots, \overset{1}{m}_{n_1}, \quad \overset{2}{m}_1, \ldots, \overset{2}{m}_{n_2}, \ldots, \overset{k}{m}_1, \ldots, \overset{k}{m}_{n_k}\}$$

where the set of the first n_1 elements form a basis of M_1 and in general the set of the first $n_1 + n_2 + \cdots + n_i$ elements are a basis of M_i, for $i = 1, 2, \ldots, k$. We have now

$$\begin{array}{lllll}
\overset{1}{m}_1 r = & \overset{1}{f}_{11}\overset{1}{m}_1 + \cdots + \overset{1}{f}_{1n_1}\overset{1}{m}_{n_1} & & & \\
\cdot & \cdot \quad \cdots \quad \cdot & & & \\
\overset{1}{m}_{n_1} r = & \overset{1}{f}_{n_1 1}\overset{1}{m}_1 + \cdots + \overset{1}{f}_{n_1 n_1}\overset{1}{m}_{n_1} & & & \\
\overset{2}{m}_1 r = & * & + \overset{2}{f}_{11}\overset{2}{m}_1 + \cdots + \overset{2}{f}_{1n_2}\overset{2}{m}_{n_2} & & \\
\cdot & \cdot & \cdot \quad \cdot \quad \cdot & & \\
\overset{2}{m}_{n_2} r = & * & + \overset{2}{f}_{n_2 1}\overset{2}{m}_1 + \cdots + \overset{2}{f}_{n_2 n_2}\overset{2}{m}_{n_2} & & \\
\vdots & & & \ddots & \\
\overset{k}{m}_1 r = & * & & * & + \overset{k}{f}_{11}\overset{k}{m}_1 + \cdots + \overset{k}{f}_{1n_k}\overset{k}{m}_{n_k} \\
\cdot & \cdot & \cdot \quad \cdot & \cdot \quad \cdot & \cdot \quad \cdot \quad \cdots \quad \cdot \\
\overset{k}{m}_{n_k} r = & * & & * & + \overset{k}{f}_{n_k 1}\overset{k}{m}_1 + \cdots + \overset{k}{f}_{n_k n_k}\overset{k}{m}_{n_k}.
\end{array}$$

In each line an asterisk indicates a linear combination over F of earlier basis elements. This gives the matrix representation

$$\text{(4.12)} \qquad \mu(r) = \begin{pmatrix} \mu_1(r) & & & \\ & \mu_2(r) & & \\ & & \ddots & \\ * & & & \mu_k(r) \end{pmatrix}$$

where $\mu_i(r)$ stands for the $n_i \times n_i$ matrix

$$\begin{pmatrix} f_{11}^i & \cdots & f_{1n_i}^i \\ \cdot & \cdots & \cdot \\ f_{n_i1}^i & \cdots & f_{n_in_i}^i \end{pmatrix}.$$

This is the matrix for $\mu_i(r)$ corresponding to the basis for M_i modulo M_{i-1}, that is, for the module M_i/M_{i-1}.

A basis for M adapted to the decomposition (4.9) will give for μ a matrix of the form (4.12) with the asterisk replaced by zero. In this case the diagonal blocks are indecomposable. They may be reducible.

Let χ^μ be the character (Definition 2.1) of the representation μ. From the matrix (4.12) we get

(4.13) $$\chi^\mu(r) = \chi^{\mu_1}(r) + \cdots + \chi^{\mu_k}(r).$$

This gives:

(4.14) **Lemma.** *The character of a representation is the sum of the characters of its irreducible constituents.*

5. The Regular Representation

Let R be a ring. With respect to its addition R is a module, and with the ring multiplication, it is even an R-module. Hence R is its own representation module. The corresponding representation ρ is called the right regular representation. We have

$$\forall x \in R : x\rho(r) = xr, \qquad r \in R$$

The admissable submodules of the regular representation are those subrings $R_i \subset R$ for which $R_i x \subset R_i$, $\forall x \in R$. Thus the R_i are *right ideals* of the ring R. Furthermore, any *minimal right ideal* of R is a representation module for an irreducible representation of R.

The importance of the regular representation is shown by:

(5.1) **Lemma.** *Every irreducible representation μ of a ring R is equivalent to a constituent of the regular representation.*

Proof. Let M be the representation module for the irreducible representation μ. Since μ is not the zero representation, $MR \neq 0$ and $\exists m \in M$ so that $mR \neq 0$. Hence $mR = M$, since mR is a submodule of the irreducible module M.

Let π be the mapping: $R \to M$ given by $r\pi = mr$, $\forall r \in R$. Since

$$(r + r')\pi = m(r + r') = mr + mr' = r\pi + r'\pi$$

and

$$(rr')\pi = mrr' = r\pi r',$$

therefore π is an operator homomorphism of R as a module onto M. Let K be the kernel of π. By the first isomorphism theorem of group theory, $R/K \cong M$, therefore the representation σ corresponding to R/K is equivalent to the representation μ corresponding to M. Since R/K is a bottom constituent of ρ the lemma is proved.

Remark. A similar statement for indecomposable representations is not true in general. Under certain circumstances a ring can have indecomposable representations of arbitrarily high degree, whereas the degree of a component of the regular representation is necessarily bounded by the degree of the latter (see [13] and Exercise 4 below).

EXERCISES

1. The symmetric group S_3 of order 6 is generated by all products of the permutations

$$a = \begin{pmatrix} 1 & 2 & 3 \\ 2 & 1 & 3 \end{pmatrix} \quad \text{and} \quad b = \begin{pmatrix} 1 & 2 & 3 \\ 3 & 1 & 2 \end{pmatrix},$$

written concisely as

$$\begin{pmatrix} j \\ ja \end{pmatrix} \quad \text{and} \quad \begin{pmatrix} j \\ jb \end{pmatrix},$$

respectively, $j = 1, 2, 3$. Let V be a vector space with a basis X_1, X_2, X_3 over the real numbers F. Show that defining

$X_j s = X_{js}$, $s \in S_3$, with extension to V by linearity, makes V into an F-S_3 module. Find:

(i) the matrix representation ν by using the given basis,

(ii) the matrix representation μ using the basis

$$Y_1 = X_1 + X_2 + X_3\,,\ Y_2 = X_1 - X_2\,,\ Y_3 = X_1 - X_3\,.$$

Show that the components of μ are indecomposable. Find the matrix T such that $\mu(s) = T\nu(s)T^{-1}$, $\forall s \in S_3$ (cf. example at the end of the Introduction).

2. A group can be specified by giving a set of generators and defining relations. Thus given the generators a, b and the relation $a^2 = b^2 = 1$, $ab = ba$ the set $G = \{1, a, b, ab\}$ of distinct elements is closed under multiplication, i.e., the product of any two of them can be expressed as an element in the set (using the relations if necessary). Here G is a group of order 4, *the Klein 4-group*. Show that the regular matrix representation of G (see Observation 1 of the Introduction) can be given by

$$\rho(a) = \begin{pmatrix} J & \check{0} \\ \check{0} & J \end{pmatrix}, \qquad \rho(b) = \begin{pmatrix} \check{0} & I \\ I & \check{0} \end{pmatrix}$$

where

$$J = \begin{pmatrix} 0 & 1 \\ 1 & 0 \end{pmatrix}, \qquad I = \begin{pmatrix} 1 & 0 \\ 0 & 1 \end{pmatrix},$$

and $\check{0}$ is the 2×2 zero matrix. If

$$A = \begin{pmatrix} x & y \\ y & x \end{pmatrix}, \qquad B = \begin{pmatrix} z & w \\ w & z \end{pmatrix},$$

show that the group matrix is $\begin{pmatrix} A & B \\ B & A \end{pmatrix}$ and that

$$T\begin{pmatrix} A & B \\ B & A \end{pmatrix}T^{-1} =$$

$$\begin{pmatrix} x+y+z+w & 0 & 0 & 0 \\ 0 & x+y-z-w & 0 & 0 \\ 0 & 0 & x-y+z-w & 0 \\ 0 & 0 & 0 & x-y-z+w \end{pmatrix}$$

where

$$T = \begin{pmatrix} 1 & 1 & 1 & 1 \\ 1 & 1 & -1 & -1 \\ 1 & -1 & 1 & -1 \\ 1 & -1 & -1 & 1 \end{pmatrix}.$$

3. Let $C = \{a + ib : a, \; b \in F, \text{ the real field}\}$ be the set of complex numbers. Using C as its own F-C module, with basis $\{1, i\}$ over F, find the regular matrix representation $\rho(a + ib)$.

4. Let G be the Klein 4-group (Exercise 2). Let $F = \{0, 1\}$ be the field of two elements (0, 1 are integers and addition and multiplication is performed modulo 2). The set

$$R = \left\{\sum_{i=1}^{4} f_i g_i : f_i \in F, \quad g_i \in G\right\}$$

of formal sums, added and multiplied like algebraic expressions together with the group multiplication for the product of the g_i, form a ring or algebra: *the group algebra of G over F*. Using R as its own F-R module, with basis $\{1, a, b, ab\}$, we can find the regular matrix representation ρ of R. Restricted to G, ρ gives the regular representation of G. Its degree is 4. Now let $V_5 = \{X = (x_1, x_2, x_3, x_4, x_5) : x_i \in F\}$ be a vector space over F. Show that defining: $Xa = (x_1, x_2, x_2 + x_3, x_4, x_4 + x_5)$, $Xb = (x_1, \; x_2, \; x_1 + x_2 + x_3, x_4, \; x_2 + x_4 + x_5)$, and extending the definition to R by linearity makes V_5 into an F-R module which is absolutely indecomposable. Then the corresponding matrix representation is an absolutely indecomposable representation of degree 5 and cannot be a component of the regular representation.

5. Let G be a group and S a subgroup. Consider the right coset decomposition: $G = Sx_1 + Sx_2 + \cdots + Sx_n$. Let $L = \{x_1, ..., x_n\}$, $x_1 = 1$, be a fixed set of representatives. Note that $\forall g \in G$, g can be written uniquely: $g = g^* g_*$, $g^* \in S$, and $g_* \in L$. Observe that $(sg)^* = sg^*$, $(sg)_* = g_*$, and that $(xg)_*$ spans L as x runs through

L. Let M be an F-S module. Let $M^G = \{\sigma : L \to M\}$ be the set of all mappings of L into M. Define

$$\left.\begin{array}{rl} \sigma + \tau: & x(\sigma + \tau) = x\sigma + x\tau \\ f\sigma, f \in F: & x(f\sigma) = f(x\sigma) \\ \sigma g, g \in G: & (xg)_*(\sigma g) = (x\sigma)(xg)^* \end{array}\right\} \quad x \in L$$

Show that M^G is an F-G module and that M^G restricted to S contains a submodule $M' \cong M$ under the correspondence: $M^G \ni \sigma \leftrightarrow m \in M$, if and only if $1\sigma = m$, $x\sigma = 0$, $x \neq 1$. In this way a representation of S *induces* a representation of G.

CHAPTER II

Representation Theory of Rings with Identity

6. Some Fundamental Lemmas

In this section we prove a number of results that will be needed in the development of the theory. The lemmas concern components of a representation and particularly of the regular representation.

Let us recall that a module M with operators is a direct sum of its admissable submodules $M_1, \ldots, M_k$, written

$$M = M_1 \oplus \cdots \oplus M_k, \tag{6.1}$$

if $\forall m \in M$, there is a *unique* relation

$$m = m_1 + \cdots + m_k, \qquad m_i \in M_i. \tag{6.2}$$

By virtue of (6.2) there is associated with a given decomposition (6.1) a number of decomposition operators $\delta_i : M \rightarrow M_i$ defined by $m\delta_i = m_i$. Now if $m' \in M$, let $m' = m_1' + \cdots + m_k'$. Then

$$m + m' = (m_1 + m_1') + \cdots + (m_k + m_k').$$

Also

$$mr = m_1 r + \cdots + m_k r,$$

if r is an operator. Moreover, both expressions are unique. Now we see that

$$(m + m')\delta_i = m_i + m_i' = m\delta_i + m'\delta_i$$

and

$$mr\delta_i = m_i r = m\delta_i r$$

which shows that the δ_i are operator homomorphisms of M into itself, i.e., *operator endomorphisms* of M.

Furthermore, since $m_i = 0 + \cdots + m_i + \cdots + 0$ the uniqueness of (6.2) gives

$$m_i\delta_i = m_i\,, \quad \forall\, m_i \in M_i\,, \text{ and } m_j\delta_i = 0, \quad \forall\, m_j \in M_j\,, \quad \text{if } j \neq i.$$

These results show that

$$\delta_i^2 = \delta_i\,, \qquad \delta_i\delta_j = 0, \qquad \text{if} \quad i \neq j.$$

Finally, from (6.2) we have

$$m = m\delta_1 + \cdots + m\delta_k = m(\delta_1 + \cdots + \delta_k)$$

so that

$$\delta_1 + \cdots + \delta_k = 1, \qquad \text{the identity map:} \quad M \to M.$$

(6.3) **Lemma.** *Let R be a ring with an identity and let the right ideal R_1 be a direct summand. If ρ is an operator homomorphism of R_1 into an R-module N and if ν is an operator homomorphism of another R-module M* **onto** *N, then there exists an operator homomorphism $\tau : R_1 \to M$, such that $\rho = \tau\nu$.*

In other words, there exists a τ which makes the following diagram commutative:

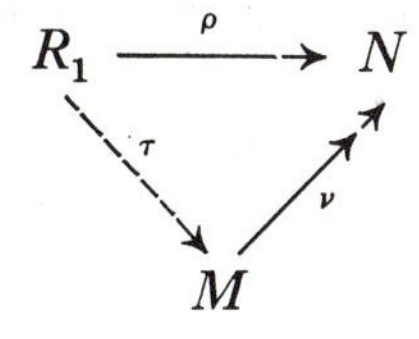

Proof. Let δ be the decomposition operator of the decomposition

$$R = R_1 \oplus \cdots$$

which maps R onto R_1. Let $(1\delta)\rho = n \in N$, 1 the identity of R. Since ν is onto, $\exists m \in M$ such that $m\nu = n$.

Now let τ be the mapping $R_1 \to M$ defined by $x\tau = mx$, $\forall x \in R_1$. Since

$$(xr)\tau = m(xr) = (mx)r = (x\tau)r, \qquad r \in R,$$

and

$$(x + y)\tau = m(x + y) = mx + my = x\tau + y\tau, \qquad x, y \in R_1,$$

therefore τ is an operator homomorphism of R_1 into M. Moreover, $\forall x \in R_1$, $x\delta = x$, and

$$x\rho = (x\delta)\rho = ((1x)\delta)\rho = ((1\delta)x)\rho = (1\delta)\rho x = nx = m\nu x$$

$$= (mx)\nu = (x\tau)\nu = x(\tau\nu).$$

$$\therefore \quad \rho = \tau\nu.$$

(6.4) **Corollary.** *If $\check{N}$ is an admissable submodule of an R-module $\check{M}$ and if $\rho : R_1 \to \check{M}/\check{N}$, then $\exists\tau : R_1 \to \check{M}$.*

Proof. Take $\check{M}/\check{N}$ as the N of the lemma and ν as the natural homomorphism $\check{M} \twoheadrightarrow \check{M}/\check{N}$.

(6.5) **Corollary.** *If $\check{N}$ of Corollary 6.4 is a* ***unique maximal submodule*** *of $\check{M}$ then $\tau : R_1 \twoheadrightarrow \check{M}$; that is, τ is onto.*

Proof. $R_1\tau$ is an R-module and if τ is not onto: $R_1\tau \subset \check{N}$.

$$\therefore \quad R_1\tau\nu = 0 = R_1\rho; \qquad \text{since } \rho \neq 0 \text{ this is impossible.}$$

$$\therefore \quad \tau \text{ is onto.}$$

(6.6) **Corollary.** *Let $\nu : M \twoheadrightarrow R_1$ be an operator homomorphism of an R-module M* ***onto*** *R_1, then $M = M_1 \oplus M_2$, where $M_1 \cong R_1$ and M_2 is the kernel of ν.*

Proof. Let ρ of Lemma 6.3 be the identity map $i : R_1 \to R_1$ and hence determine $\tau : R_1 \to M$ such that $\tau\nu = i$:

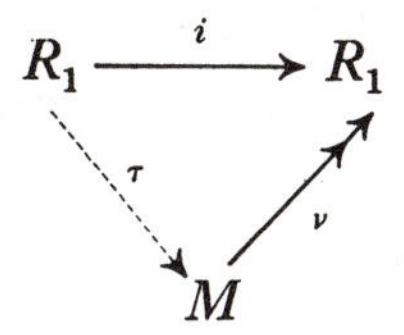

Let $R_1\tau = M_1$ and let M_2 be the kernel of ν. Since $\tau\nu = i$ is an isomorphism, therefore τ is an isomorphism and $R_1 \cong M_1$. If $m' \in M_1 \cap M_2$, then $\exists r_1 \in R_1$ such that $m' = r_1\tau$. But $0. = m'\nu = r_1\tau\nu = r_1$, so that $m' = 0$. Thus $M_1 \cap M_2 = 0$. Again let $m \in M$ and put $m\nu = r_1 \in R_1$. Then

$$m = r_1\tau + (m - r_1\tau), \qquad \text{and} \qquad r_1\tau \in M_1 .$$

Now $(m - r_1\tau)\nu = m\nu - r_1\tau\nu = r_1 - r_1 = 0$, so that $(m - r_1\tau) \in M_2$. Hence

$$M = M_1 \oplus M_2 .$$

It will be convenient for later use to have Corollary 6.6 reformulated in matrix terms as:

(6.6′) **Corollary.** *Let*

$$\mu(x) = \begin{pmatrix} \mu_1(x) & 0 \\ C(x) & \rho_1(x) \end{pmatrix}$$

be a reducible matrix representation of the ring R. If the bottom constituent $\rho_1(x)$ is a component of the regular representation $\rho(x)$, then there is a matrix

$$T = \begin{pmatrix} I & 0 \\ H & I \end{pmatrix}$$

such that

$$T^{-1}\mu(x)T = \begin{pmatrix} \mu_1(x) & 0 \\ 0 & \rho_1(x) \end{pmatrix}.$$

Proof. Let M be the representation module, with basis $\{m_1, m_2, \ldots, m_{f_1}, m_{f_1+1}, \ldots, m_{f_1+f_2}\}$, which affords the representation μ. The first f_1 elements are a basis of the representation submodule M_1 which gives the representation μ_1. Then M/M_1 affords $\rho_1(x)$, which by hypothesis is afforded by a direct summand R_1 of R. Thus

$$M \twoheadrightarrow M/M_1 \cong R_1$$

and Corollary 6.6 gives

$$M = M_1 \oplus M_1', \qquad M_1' \cong R_1. \tag{6.7}$$

For each m_j, $f_1 < j \leqslant f_1 + f_2$ we have by (6.7)

$$m_j = m_{1j} + m_j', \qquad m_{1j} \in M_1,$$

or, using a basis of M_1,

$$m_j = \sum_{k=1}^{f_1} h_{jk} m_k + m_j', \qquad f_1 < j \leqslant f_1 + f_2. \tag{6.8}$$

Then the set $\{m_1, \ldots, m_{f_1}\ m_{f_1+1}', \ldots, m_{f_1+f_2}'\}$ is a new basis of M adapted to the decomposition (6.7). It affords a representation μ', similar to μ, of the form required. Since the bases are connected by $m_i = m_i$, $1 \leqslant i \leqslant f_1$, and Eqs. (6.8), the transforming matrix T also has the required form.

(6.9) **Lemma.** *Let M be an indecomposable representation module of the ring R which possesses a finite composition series of admissable submodules: $M = M_0 \supset M_1 \supset \cdots \supset M_l = 0$, M_i maximal in M_{i-1}. Let θ be an operator endomorphism of M, then either θ is an automorphism of M* **onto** *M or θ is nilpotent and $\theta^l = 0$.*

This result is a special case of Fitting's lemma in group theory, but we give a direct proof.

Proof. The existence of a finite composition series for M is equivalent to the double chain condition. This means that sequences of admissable submodules of the following two types:

$$\text{(a)} \quad M_1 \subset M_2 \subset M_3 \subset \cdots$$
$$\text{(b)} \quad M_1 \supset M_2 \supset M_3 \supset \cdots$$

contain only a finite number of distinct terms. Let

$$N_j = \{m : m \in M,\ m\theta^j = 0\}.$$

Then $N_1 \subset N_2 \subset \cdots$ and by (a) $\exists s$ such that $N_s = N_{s+1}$. Thus $m\theta^{s+1} = 0 \Rightarrow m\theta^s = 0$.

Now θ induces an automorphism of $M/N_s = \bar{M}$, for if $\bar{m} \in \bar{M}$:

$$\bar{m}\theta = 0 \Rightarrow \overline{m\theta} = 0 \Rightarrow m\theta \in N_s \Rightarrow (m\theta)\theta^s = m\theta^{s+1} = 0$$
$$\Rightarrow m\theta^s = 0 \Rightarrow m \in N_s \Rightarrow \bar{m} = 0.$$

Moreover, θ is *onto* $\bar{M}$, for (b) holding in M also holds in $\bar{M}$ and so $\bar{M} \supset \bar{M}\theta \supset \cdots \supset \bar{M}\theta^k = \bar{M}\theta^{k+1}$ for some k. Hence $\forall \bar{m} \in \bar{M}$, $\exists \bar{m}_1 \in \bar{M}$ such that $\bar{m}\theta^k = \bar{m}_1\theta^{k+1}$. Since θ is an automorphism this implies $\bar{m} = \bar{m}_1\theta$, so that every element of $\bar{M}$ is an image of another under θ and $\bar{M} = \bar{M}\theta = \bar{M}\theta^2 = \cdots = \bar{M}\theta^s$. But then $\forall \bar{m}$, $\exists \bar{m}_2$ such that $\bar{m} = \bar{m}_2\theta^s = \overline{m_2\theta^s}$. This implies that $m = m_2\theta^s + n$, $n \in N_s$. Therefore

$$M = M\theta^s + N_s .$$

Finally if $x \in M\theta^s \cap N_s$, we have $x\theta^s = 0$ and $x = m\theta^s$, for some $m \in M$. Therefore

$$0 = x\theta^s = m\theta^{2s},$$

and thus $m \in N_s$ giving $x = m\theta^s = 0$. Therefore

$$M = M\theta^s \oplus N_s$$

Since M is indecomposable *either* $N_s = 0$, making θ an automorphism of M onto itself, *or* $M = N_s$ in which case $M\theta^s = 0$ and so $\theta^s = 0$. As the length of a series in M, $s \leqslant l$, and hence $\theta^l = 0$. This proves the lemma.

Note. It is easy to see that the converse of Lemma 6.9 holds, viz. if every operator endomorphism of an R-module M is either an automorphism or is nilpotent, then M is indecomposable. For if $M = M_1 \oplus M_2$ then the decomposition operator δ_1 is neither nilpotent nor an automorphism, since $\delta_1^2 = \delta_1$ and, $\forall x, (x\delta_2)\delta_1 = 0$.

This result gives a useful criterion for deciding whether or not a module is decomposable.

Example. Let R be the ring of integers modulo 8. If θ is an endomorphism let $1\theta = t$. Then $\forall s \in R, s\theta = (1 + 1 + \cdots + 1)\theta = (1\theta)s = ts$. Now if $t \not\equiv 0$ (Mod 2), θ is an automorphism, and if $t \equiv 0$ (Mod 2), θ is nilpotent. Hence R is indecomposable.

EXERCISE

Let A be the group algebra of the Klein 4-group over $F = \{0, 1\}$ (see Exercise 4, Chapter I). Show that the ideal generated by $(1 + a)$ is indecomposable but reducible.

7. The Principal Indecomposable Representations

Let R be a ring. An indecomposable component of the regular representation of R is called a principal indecomposable representation. It is the representation provided by a minimal right ideal R_1 which is a direct summand: $R = R_1 \oplus \tilde{R}$ ($\tilde{R}$ is a right ideal).

We shall assume that the double chain condition holds for right ideals of R:

every ascending chain of right ideals

$$R_1 \subset R_2 \subset R_3 \subset \cdots$$

and every descending chain of right ideals

$$R_1 \supset R_2 \supset R_3 \supset \cdots$$

contains only a finite number of distinct ideals.

These are equivalent, respectively, to the condition that every set of right ideals of R has a maximal and a minimal right ideal. Under these conditions R will have a Remak decomposition:

$$R = R_1 \oplus R_2 \oplus \cdots \oplus R_k$$

in which the R_i are indecomposable right ideals. We have encountered this case before [in (4.9)] for F-R modules, where the finite F-basis condition implies the double chain condition for admissable submodules. We rephrase the result there (Theorem 4.11) as:

(7.1) **Theorem.** *A ring R, with the double chain condition on right ideals, has a finite number k of principal indecomposable representations. They are unique up to equivalence.*

(7.2) **Theorem.** *Let R_i, $i = 1, \ldots, k$, be indecomposable right ideals of a ring R and let*

$$R = R_1 \oplus R_2 \oplus \cdots \oplus R_k .$$

Then

(a) There is a *unique maximal right ideal* $R_i' \subset R_i$, $i = 1, \ldots, k$.

(b) $R_i \cong R_j$ if and only if $R_i/R_i' \cong R_j/R_j'$.

(c) Every irreducible representation of R is provided by a representation module R_i/R_i' for some i.

(d) There are exactly as many inequivalent irreducible representations of R as there are inequivalent principal indecomposable representations.

Proof. (d) Clearly (b) and (c) $\Rightarrow$ (d).

(a) Let X be one of the R_i. Let $S \neq X$ be a *maximal right ideal of* R in X, and $T \neq X$ be an arbitrary right ideal of R in X. Suppose $T \not\subset S$, then $S \subset T + S$, and since S is maximal: $X = S + T$. Consider the decomposition operator $\delta : R \twoheadrightarrow X$ (Section 6). Let

$$1\delta = e' = s + t, \qquad s \in S, \quad t \in T.$$

Define

$$i', \sigma, \tau : X \to X \qquad \text{by} \qquad xi' = x, \qquad x\sigma = sx, \qquad x\tau = tx$$

Since

$$x\delta = x, \qquad \forall x \in X : xi' = x = x\delta = (1x)\delta = (1\delta)x$$
$$= (s + t)x = x\sigma + x\tau.$$
$$\therefore \quad i' = \sigma + \tau.$$

Moreover $\sigma \neq 0$, otherwise $X = Xi' = X\tau = tX \subset T$. Similarly $\tau \neq 0$. Now $(xr)\sigma = s(xr) = (sx)r = x\sigma r$, $\forall r \in R$. Hence, since $X\sigma \subset S$, σ and likewise τ are operator endomorphisms of X *into* X. By Lemma 6.9 they are nilpotent. Thus $\exists m, n > 1$ such that $\sigma^m = 0$, $\sigma^{m-1} \neq 0$, $\tau^n = 0$, $\tau^{n-1} \neq 0$. But then

$$\sigma^{m-1} = i'\sigma^{m-1} = (\sigma + \tau)\sigma^{m-1} = \tau\sigma^{m-1} = \tau^2\sigma^{m-1} = \cdots$$
$$= \tau^n\sigma^{m-1} = 0$$

giving a contradiction. Therefore $T \subset S$, so that $S = R_i'$ is a unique maximal right ideal in $X = R_i$.

(b) Assume $R_i/R_i' \cong R_j/R_j'$. The following scheme of mappings is evident:

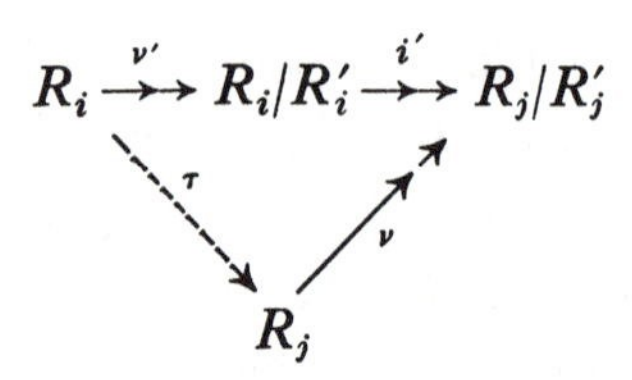

By Corollaries 6.4 and 6.5 with $\rho = \nu' i'$, $\exists \tau : R_i \twoheadrightarrow R_j$. Then by Corollary 6.6, since R_i is indecomposable, $R_i \cong R_j$. On the other hand, if τ is the isomorphism $R_i \cong R_j$, then $R_i'\tau = R_j'$ which leads to the isomorphism $R_i/R_i' \cong R_j/R_j'$.

(c) Let M be an irredicible R-module. Since $MR \neq 0$, $\exists i$ such that $MR_i \neq 0$. Then $\exists m \in M$ such that $mR_i \neq 0$. Since M is irreducible and mR_i is a submodule, therefore $mR_i = M$. Let $\tau : R_i \to M$ be defined by $r\tau = mr$, $\forall r \in R_i$. Let $\tilde{R}_i$ be the kernel of τ. Then $R_i/\tilde{R}_i \cong M$. Now M is irreducible, hence $\tilde{R}_i$ is maximal and so by (a) $\tilde{R}_i = R'_i$. Thus $R_i/R'_i \cong M$. The proof is now complete.

Example. Let R be the ring of integers modulo 8. Now R is indecomposable (Example, Section 6) and by Theorem 7.2 has a unique maximal ideal R'. Here $R' = \{0, 2, 4, 6\}$. The principal indecomposable representation maps each element $a \in R$ onto the endomorphism $\rho(a)$ given by

$$x\rho(a) = xa, \qquad \forall x \in R.$$

The irreducible representation σ provided by R/R' is given by

$$\bar{x}\sigma(a) = \overline{xa}, \qquad \forall \bar{x} \in R/R'.$$

The study of principal indecomposable representations will be resumed in Section 12.

8. The Radical of a Ring

Let R be a ring with an identity in which the double chain condition holds for right ideals. Theorems 7.1 and 7.2(c) show that there are only a finite number of distinct irreducible representations $\rho_1, \rho_2, \ldots, \rho_s$ of R.

(8.1) **Definition.** *The radical N of R is the set of all elements $r \in R$ for which $\rho_i(r) = 0$, for every irreducible representation ρ_i of R:*

$$N = \{r : r \in R, \quad \rho_i(r) = 0, \quad i = 1, 2, \ldots, s\}.$$

(8.2) **Lemma.** *N is a two-sided ideal of R.*

Proof. If $n_1, n_2, n \in N$, $r \in R$, and ρ_i is any irreducible representation, then $0 = \rho_i(n_1 \pm n_2) = \rho_i(rn) = \rho_i(nr)$. Hence $n_1 \pm n_2 \in N$ $nr \in N$, $rn \in N$, showing that N is a two-sided ideal of R.

Let us recall that the product of two subsets S_1 and S_2 of a ring R is the set of all finite sums $\Sigma\, s_1 s_2$, $s_1 \in S_1$, $s_2 \in S_2$. Then, for example, N^l as a set is the set of all finite sums $\Sigma\, n_1 n_2 \cdots n_l$, $n_i \in N$. Under the addition and multiplication of the ring this set forms a two-sided ideal which we denote by N^l.

(8.3) **Lemma.** *N is nilpotent, that is there is a positive integer l so that $N^l = 0$.*

Proof. Let $R_1 \supset R_2 \supset \cdots \supset R_l = 0$ be a composition series for R. Now $\forall n \in N$, $R_i n \subset R_{i+1}$, since n must be represented by zero in the irreducible representation afforded by the module R_i/R_{i+1}. Thus

$$Rn_1 n_2 \cdots n_l \subset R_l = 0,$$

and since $1 \in R$,

$$1 n_1 n_2 \cdots n_l = 0,$$

that is, $N^l = 0$.

(8.4) **Theorem.** *Let $R = R_1 \oplus \cdots \oplus R_k$, R_i indecomposable and let R_i' be the unique maximal* [Theorem 7.2(a)] *subideal of R_i, then*

$$N = R_1' \oplus \cdots \oplus R_k'$$

where N is the radical of R.

Proof. (a) Let $n \in N$. Now $R_i n \subseteq R_i'$, since n must be represented by zero through the irreducible representation module R_i/R_i'. Hence

$$Rn = R_1 n + \cdots + R_k n \subseteq R_1' \oplus \cdots \oplus R_k'.$$

Since $1 \in R$,

$$\therefore \quad n \in R_1' \oplus \cdots \oplus R_k', \qquad \text{and hence} \qquad N \subseteq R_1' \oplus \cdots \oplus R_k'.$$

(b) Since the R_i are right ideals $R_i R_j' \subseteq R_i$. Now if we prove that $R_i R_j' \subseteq R_i'$, $\forall i, j$ then $\rho_i(R_j') = 0$ for all irreducible representations so that $R'_j \subset N$, $\forall j$. Hence $N \supset R'_1 + \cdots + R'_k$. Together with the final inequality of part (a) this would give

$$N = R'_1 \oplus \cdots \oplus R'_k .$$

To complete the proof we must show that $R_i R'_j \neq R_i$; for then, since $R_i R'_j$ is a proper right ideal of R_i and since R'_i is the unique maximal ideal in R_i, we get $R_i R'_j \subseteq R'_i$.

Suppose $R_i R'_j = R_i$. Then $\exists z \in R_i$ such that $zR'_j = R_i$ (otherwise $xR'_j \subset R_i$, $\forall x \in R_i$ and since the xR'_j are right ideals and R'_i is unique maximal: $xR'_j \subset R'_i$ so that $R_i R'_j \subseteq R'_i \neq R_i$). But then also $zR_j = R_i$. Hence $\forall x \in R_j$, $\exists x' \in R'_j$ such that $zx = zx'$ or $z(x - x') = 0$. Now remark that $A = \{a : a \in R_j, za = 0\}$, the set of right annihilators of z, is a right ideal in R_j and $A \neq R_j$ since $zR_j = R_i \neq 0$. Therefore $A \subset R'_j$, since R'_j is unique maximal in R_j. Then $x - x' \in R'_j$; but $x' \in R'_j$, therefore $x \in R'_j$, therefore $R_j \subseteq R'_j$ giving a contradiction. This establishes the lemma.

(8.5) **Lemma.** *N contains all nilpotent right and left ideals.*

Proof. (a) Let D be a right ideal of R and suppose $D^l = 0$. If $d \notin N$ there is some irreducible representation ρ_i such that $\rho_i(d) \neq 0$ and this means that in the associated representation module which must occur as some R_i/R'_i [Theorem 7.2(c)] we must have $R_i d \nsubseteq R'_i$. Then $R_i D \nsubseteq R'_i$. Since $R_i D$ is a right ideal and R'_i is unique and maximal: $R_i D = R_i$. But then $R_i = R_i D = R_i D^l = 0$, which is impossible. Therefore $D \subseteq N$.

(b) Let L be a nilpotent left ideal. Remark that LR is a nilpotent right ideal so that by (a) $LR \subseteq N$. Since R has an identity $L \subseteq LR \subseteq N$ and the proof is complete.

Remark. Since N is nilpotent every right or left ideal in N is nilpotent. Hence from Lemma 8.5 the radical can be characterised as the largest nilpotent right or left ideal of a ring.

Example. Let

$$R = \left\{ \begin{pmatrix} a & 0 \\ c & b \end{pmatrix}, \quad a, b, c \text{ real numbers} \right\}.$$

Then

$$A = \left\{\begin{pmatrix} a & 0 \\ 0 & 0 \end{pmatrix}\right\} \qquad \text{and} \qquad B = \left\{\begin{pmatrix} 0 & 0 \\ c & b \end{pmatrix}\right\}$$

are right ideals and

$$R = A \oplus B, \qquad A, B \text{ indecomposable.}$$

A is irreducible but

$$B \supset C = \left\{\begin{pmatrix} 0 & 0 \\ c & 0 \end{pmatrix}\right\}.$$

C is the maximal right ideal of R in B. It is two-sided and nilpotent and is the radical. As representation modules, A gives the irreducible representation

$$\rho_1\left(\begin{pmatrix} a & 0 \\ c & b \end{pmatrix}\right) = a,$$

and B/C the irreducible representation

$$\rho_2\left(\begin{pmatrix} a & 0 \\ c & b \end{pmatrix}\right) = b.$$

There are no other irreducible representations.

9. Semisimple Rings

(9.1) **Definition.** *A ring is semisimple if its radical N is zero.*

From Theorem 8.4 we see that R is semisimple if and only if the indecomposable components of R are irreducible, or in other words, if R is completely reducible. This can be stated as:

(9.2) **Theorem.** *R is semisimple if and only if its regular representation is completely reducible.*

(9.3) **Theorem.** *Every representation μ of a semisimple ring R is completely reducible.*

Proof. Let M be the representation module for μ and let

$$M = M_1 \oplus \cdots \oplus M_s, \qquad M_i \text{ indecomposable.}$$

We must show that the M_i are also irreducible. Now since R has an identity by assumption and M is unitary (condition 9, Definition 3.1) $M_iR \neq 0$, $\forall i$. Also since R is semisimple $R = \Sigma_{i=1}^{k} \oplus R_i$, R_i irreducible right ideals. Let M_i' be a maximal representation submodule of M_i. Then M_i/M_i' is irreducible and by Theorem 7.2(c) there is an R_j such that $M_i/M_i' \simeq R_j$ ($R_j' = 0$ by the semisimplicity). Consider the scheme of mappings:

$$M_i \xrightarrow{\nu} M_i/M_i' \xrightarrow{i} R_j$$

where ν is the natural homomorphism and i is the assumed isomorphism. Then νi maps M_i *onto* R_j so that by Corollary 6.6, since M_i is indecomposable, $M_i \simeq R_j$. This implies that ν is an isomorphism and hence $M_i' = 0$ so that the lemma is proved.

(9.4) **Definition.** *A ring is simple if it has no two-sided ideals other than itself and the zero ideal.*

(9.5) **Lemma.** *R simple $\Rightarrow$ R semisimple.*

Proof. Let N be the radical of R. Since N is a two-sided ideal either $N = 0$ or $N = R$. If $N = R$ the nilpotentcy of N gives $0 = N^l = R^l$ which is impossible, since by assumption R has an identity. Hence $N = 0$ and R is semisimple.

Let us recall that if B is a two-sided ideal of a ring R then the elements taken modulo B form a ring, the difference ring, denoted by $R - B$. Its elements are the equivalence classes of R modulo B. If $\lfloor r \rfloor$ denotes the equivalence class of r, addition and multiplication are defined:

$$\lfloor r \rfloor + \lfloor r' \rfloor = \lfloor r + r' \rfloor, \qquad \lfloor r \rfloor \lfloor r' \rfloor = \lfloor rr' \rfloor$$

It is easy to verify that these class operations are independent of the class representatives and make $R - B$ into a ring. Moreover there is a natural homomorphism $\nu : r \rightarrow \lfloor r \rfloor$, of R onto $R - B$.

(9.6) **Theorem.** *$R - N$ is a semisimple ring.*

Proof. The existence of identity and of a composition series for $R - N$ follows from their existence for R. Now if ρ is any representation of R for which $\rho(N) = 0$ we see that a representation $\hat{\rho}$ of $R - N$ is well defined by

$$\hat{\rho}(\lfloor r \rfloor) = \rho(r).$$

Conversely, given $\hat{\rho}$, this relation defines ρ. (If M is the module for ρ: $m\rho(r) = mr = m(r + n)$, $\forall n \in N$, $\therefore$ $MN = 0$.)

In this way there is a one-to-one correspondence between the irreducible representations $\rho_1, \rho_2, ..., \rho_s$ of R and the corresponding irreducible representations $\hat{\rho}_1, \hat{\rho}_2, ..., \hat{\rho}_s$ of $R - N$. If now $\lfloor r \rfloor \in R - N$ and $\hat{\rho}_i(\lfloor r \rfloor) = 0$, $\forall i$, then $\rho_i(r) = 0$, $\forall i$, and $r \in N$ so that $\lfloor r \rfloor = 0$. Hence if $\hat{N}$ denote the radical of $R - N$, $\hat{N} = 0$, and $R - N$ is semisimple.

10. The Wedderburn Structure Theorems for Semisimple Rings

(10.1) **Lemma.** *A ring with identity which is the direct sum of isomorphic minimal right ideals is simple.*

Proof. Let $R = R_1 \oplus \cdots \oplus R_l$, R_i irreducible, $R_i \overset{\sigma_i}{\cong} R_1$, $\forall i$. Let B be a two-sided ideal of R. If $B \neq 0$, since $1 \in R$, $0 \neq RB = (R_1 \oplus \cdots \oplus R_l)B = R_1B + \cdots + R_lB$. Hence one of the summands, say R_1B, $\neq 0$. Since R_1 is a minimal right ideal $R_1B = R_1 \subset B$. But now $R_jB \neq 0$, otherwise $0 = R_jB\sigma_j = R_j\sigma_jB = R_1B = R_1$, where σ_j is the operator isomorphism of $R_j \twoheadrightarrow R_1$. Therefore

$$R_jB = R_j \subseteq B,$$

$$R = \sum R_j \subseteq B.$$

$$\therefore \quad B = R, \qquad \text{and so } R \text{ is simple.}$$

Now let R be *semisimple*. Since R is completely reducible:

$$R = (R_{11} \oplus \cdots \oplus R_{1n_1}) \oplus \cdots \oplus (R_{i1} \oplus \cdots \oplus R_{in_i}) \oplus \cdots$$
$$\oplus (R_{s1} \oplus \cdots \oplus R_{sn_s}). \tag{10.2}$$

Here the summation is arranged so that the n_i minimal right ideals R_{ij} in the ith block are mutually isomorphic, for each i, and $R_{ij} \not\cong R_{tk}$ if $i \neq t$. If $1 = \Sigma_{i=1}^{s} (e_{i1} + \cdots + e_{in_i})$ then $e_{i1} + \cdots + e_{in_i}$ is the identity for the ring $A_i = R_{i1} + \cdots + R_{in_i}$. This will follow from the proof in (10.3) that the A_i are two-sided ideals. Now Lemma 10.1, using the relation (10.2), gives:

(10.3) **Theorem.** *Every semisimple ring R is the direct sum of simple rings.*

$$R = A_1 \oplus \cdots \oplus A_s\,. \tag{10.4}$$

Moreover, the A_i *are two-sided ideals and* $A_iA_j = 0,\ i \neq j$. *The decomposition is unique.*

Proof. Only the last two statements need proof. Since the R_{ij} are right ideals $A_i = R_{i1} \oplus \cdots \oplus R_{in_i}$ is a right ideal. Let $r \in R$. If $rR_{ij} \not\subset A_i$ then for some $t, k, t \neq i$: $\exists a \in R_{ij}$ such that $(ra)\,\delta_{tk} \neq 0$, where δ_{tk} is the decomposition operator mapping $R \twoheadrightarrow R_{tk}$. Now the mapping $\tau : R_{ij} \to R_{tk}$ given by $x\tau = (rx)\delta_{tk} = r\delta_{tk}x$, $x \in R_{ij}$, is a homomorphism $\neq 0$, and since R_{tk} is irreducible it is an isomorphism. This is a contradiction. Therefore $rR_{ij} \subset A_i$, $\forall r \in R$. Then A_i is a left ideal, and so a two-sided ideal. Finally, $A_iA_j \subset A_i \cap A_j = 0$. If B is any two-sided ideal, since the A_j are simple: $B \cap A_j = 0$ or A_j. This shows that the decomposition (10.4) is actually unique.

Since a simple ring is semisimple we have as a converse to Lemma 10.1 the:

(10.5) **Corollary.** *A simple ring is the direct sum of isomorphic right ideals.*

Proof. If there were more than one summand in (10.4) R would have a proper two-sided ideal. Thus there is only one summand and it is the direct sum of isomorphic minimal ideals by (10.2).

This result shows that a simple ring can have only one irreducible representation up to equivalence.

(10.6) **Theorem.** *A ring R is simple if and only if it isomorphic to a complete matrix algebra of degree m over a division ring. The number m and the division ring are uniquely determined by R.*

Proof. (a) *only if.* Since R is simple:

$$R = R_1 \oplus \cdots \oplus R_m \tag{10.7}$$

in which the R_i are irreducible right ideals and $R_i \cong R_j \forall i, j$. Let δ_i be the decomposition operator $R \twoheadrightarrow R_i$. Recall (Section 6) that

$$1 = \delta_1 + \cdots + \delta_m, \qquad \delta_i^2 = \delta_i, \qquad \delta_i \delta_j = 0, \qquad i \neq j.$$

We consider Hom(R, R), the ring of all R-endomorphisms of R as an R-module. Thus $\sigma \in \text{Hom}(R, R)$ implies that $(r + r')\sigma = r\sigma + r'\sigma$; $(rr')\sigma = (r\sigma)r'$. $\forall j$ let τ_j be the fixed isomorphism $R_1 \overset{\tau_j}{\twoheadrightarrow} R_j$. For every $\sigma \in \text{Hom}(R, R)$ define the homomorphisms $\sigma_{ij} : R_1 \to R_1$ as follows:

$$r_1 \sigma_{ij} = r_1 \tau_i \sigma \delta_j \tau_j^{-1}, \qquad \forall r_1 \in R_1 . \tag{10.8}$$

Since R_1 is irreducible all endomorphisms, if not zero, are automorphisms and have inverses and the set of all of them form a skew field. Thus to each $\sigma \in \text{Hom}(R, R)$ there corresponds an $m \times m$ matrix (σ_{ij}) with elements σ_{ij} from a skew field. Now the σ_{ij} can be prescribed arbitrarily and the corresponding σ found again, for from (10.8) we have the relation

$$\sum_j \sigma_{ij} \tau_j = \sum_j \tau_i \sigma \delta_j = \tau_i \sigma \sum_j \delta_j = \tau_i \sigma 1 = \tau_i \sigma \tag{10.9}$$

which, since $R_1 \tau_i = R_i$, defines σ on R_i, $\forall i$, if we know the σ_{ij}. But σ on the R_i gives σ on R through (10.7).

Thus there is a unique correspondence $\sigma \leftrightarrow (\sigma_{ij})$ between elements of Hom(R, R) and all $m \times m$ matrices over a certain skew field. If $\mu \leftrightarrow (\mu_{ij})$, consider the element at the intersection of the ith row and kth column of the product matrix $(\sigma_{ij})(\mu_{ij})$. It is

$$\sum_j \sigma_{ij}\mu_{jk} = \sum_j \tau_i\sigma\delta_j\tau_j^{-1}\cdot\tau_j\mu\delta_k\tau_k^{-1} = \sum_j \tau_i\sigma\delta_j\mu\delta_k\tau_k^{-1}$$

$$= \tau_i\sigma\sum_j \delta_j\mu\delta_k\tau_k^{-1} = \tau_i\sigma\mu\delta_k\tau_k^{-1}.$$

But this is the similarly placed element of the matrix corresponding to $\sigma\mu$. Therefore

$$\sigma\mu \leftrightarrow (\sigma_{ij})(\mu_{ij}).$$

Similarly

$$\sigma + \mu \leftrightarrow (\sigma_{ij}) + (\mu_{ij}).$$

$$\therefore \quad \text{Hom}\,(R, R) \cong \mathfrak{M}_m .$$

$\mathfrak{M}_m$ is the set of all $m \times m$ matrices over the skew field K, uniquely determined as automorphisms of R_1. The number m is the number of summands in (10.7) and it is unique by the Krull-Schmidt theorem. Again, if $\sigma \in \text{Hom}(R, R)$, let $1\sigma = c \in R$. Then

$$r\sigma = (1r)\sigma = (1\sigma)r = cr.$$

Conversely, given $c \in R$, $\sigma \in \text{Hom}(R, R)$ is uniquely determined by

$$r\sigma = cr,$$

which gives again

$$1\sigma = c.$$

Thus there is a one-to-one correspondence between elements $\sigma \in \text{Hom}(R, R)$ and elements $c \in R$. If $\sigma \leftrightarrow c$, $\tau \leftrightarrow t$ we have

$$1(\sigma + \tau) = 1\sigma + 1\tau = c + t$$

and

$$1\sigma\tau = (1\sigma)\tau = c\tau = (1c)\tau = (1\tau)c = tc.$$

Hence

$$\sigma + \tau \leftrightarrow c + t \qquad \text{and} \qquad \sigma\tau \leftrightarrow tc.$$

Thus R is anti-isomorphic to Hom(R, R). Taking the transposes of the matrices in $\mathfrak{M}_m$ we get $\mathfrak{M}'_m$ anti-isomorphic to $\mathfrak{M}_m$. Therefore R is isomorphic to $\mathfrak{M}'_m$. This proves (a).

(b) *if*. Let R be the set of all $m \times m$ matrices over a skew field K. Then R has the unit matrix. Moreover R has a finite K-basis of m^2 elements so that the double chain condition holds. Denote by e_{ik} the matrix with $1 \in K$ at the intersection of the ith row and kth column and zeros elsewhere. Then $\forall a \in R$ there is a unique expression

$$a = \sum_{i,j} a_{ij} e_{ij}, \qquad a_{ij} \in K.$$

Now let $B \neq 0$ be any two-sided ideal of R. Let $a \in B$. If $a \neq 0$, some $a_{rs} \neq 0$, $a_{rs} \in K$. Since B is two sided it contains

$$a' = (b_{ij} \overset{-1}{a_{rs}} e_{ir}) a e_{sj}.$$

Using the fact that $e_{mn} e_{kf} = 0$ or e_{mf} according as $n \neq k$ or $n = k$ we have $a' = b_{ij} e_{ij}$. But i and j are arbitrary and b_{ij} is an arbitrary element of K. Therefore $B = R$ and R is simple.

In this context we give the following proof of:

(10.10) **Theorem** (Frobenius-Schur theorem). *If $\rho_1, \rho_2, \ldots, \rho_s$ are the distinct representations of a ring R and if $a_1, a_2, \ldots, a_s$ are arbitrary elements of R, then there exists an element $a \in R$ so that*

$$\rho_i(a) = \rho_i(a_i), \qquad i = 1, \ldots, s.$$

Proof. It suffices to consider R as semisimple, for if $\lfloor a \rfloor \in R - N$ were found so that $\hat{\rho}_i(\lfloor a \rfloor) = \hat{\rho}_i(\lfloor a_i \rfloor)$, then since $\hat{\rho}(\lfloor b \rfloor) = \rho(b)$ (see Theorem 9.6) we would have the result in R.

Accordingly let $R = A_1 \oplus \cdots \oplus A_s$, where the A_i are *mutually annihilating two-sided ideals*, each the direct sum of minimal right ideals (Theorem 10.3). Then for each a_i:

$$a_i = \overset{i}{a_1} + \overset{i}{a_2} + \cdots + \overset{i}{a_s}, \qquad \overset{i}{a_j} \in A_j, \quad j = 1, \ldots, s.$$

Let

$$a = \overset{1}{a_1} + \overset{2}{a_2} + \cdots + \overset{s}{a_s}$$

Suppose now that the minimal right ideal $R_{i1} \subset A_i$ is the representation module for ρ_i. Since

$$r_{i1}a = r_{i1}a_i^i = r_{i1}a_i, \qquad \forall r_{i1} \in R_{i1}.$$

$$\therefore \quad \rho_i(a) = \rho_i(a_i)$$

and the proof is complete.

11. Intertwining Numbers

In this section all modules considered are F-R modules with a finite F-basis. This includes the ring R itself, as a module. Hence R is an *algebra*. This means that R has a basis consisting of a finite number of elements of R, and every element of R is a unique linear combination of these basis elements with coefficients from F. As a consequence the double chain condition holds for ideals in R. We assume again that R has an identity.

(11.1) **Definition.** *The set of all F-R homomorphisms of M into N is denoted by* $\mathrm{Hom}(M, N)$. *Every $\sigma \in \mathrm{Hom}(M, N)$ is called an intertwining of M and N.*

(11.2) **Lemma.** $\mathrm{Hom}(M, N)$ *is a module with a finite F-basis.*

Proof. For $\sigma, \tau \in \mathrm{Hom}(M, N)$ define $\sigma + \tau$:

$$m(\sigma + \tau) = m\sigma + m\tau \qquad \forall m \in M.$$

Now

$$\begin{aligned}(m + m')(\sigma + \tau) &= (m + m')\sigma + (m + m')\tau \\ &= m\sigma + m'\sigma + m\tau + m'\tau \\ &= m(\sigma + \tau) + m'(\sigma + \tau),\end{aligned}$$

so that $\sigma + \tau$ is a homomorphism. Clearly $\sigma + \tau = \tau + \sigma$. Moreover, if $f \in F$, $r \in R$:

$$\begin{aligned}fmr(\sigma + \tau) = fmr\sigma + fmr\tau = f(m\sigma)r + f(m\tau)r &= f(m\sigma + m\tau)r \\ &= f(m(\sigma + \tau))r.\end{aligned}$$

$$\therefore \quad \sigma + \tau \in \mathrm{Hom}(M, N).$$

Now define $f\sigma = \sigma f$ for $f \in F, \sigma \in \text{Hom}(M, N)$ as follows:

$$m(f\sigma) = (fm)\sigma = f(m\sigma).$$

Then

$$\begin{aligned}(m + m')f\sigma &= f(m + m')\sigma = (fm + fm')\sigma = (fm)\sigma + (fm')\sigma \\ &= m(f\sigma) + m'(f\sigma).\end{aligned}$$

$$\therefore \quad f\sigma \in \text{Hom}(M, N).$$

We have now proved that $\text{Hom}(M, N)$ is an F module.

Let $\{m_1, ..., m_l\}$ and $\{n_1, ..., n_s\}$ be bases of M and N, respectively. Let $\alpha \in \text{Hom}(M, N)$, then $m_i\alpha = \sum_{j=1}^{s} a_{ij}n_j$, $i = 1, ..., l$, $a_{ij} \in F$. Thus α corresponds to an $l \times s$ matrix $A = (a_{ij})$. If $\beta \in \text{Hom}(M, N)$ corresponds to $B = (b_{ij})$ it is clear that $\alpha = \beta$ if and only if $A = B$. Moreover every linear relation among the α subsists among the corresponding A and conversely.

$$\therefore \quad \text{rank Hom}(M, N) \leqslant ls.$$

(11.3) **Definition.** *The intertwining number of M and N, denoted by $i(M, N)$, is the* rank(*F-dimension*) *of* $\text{Hom}(M, N)$.

(11.4) **Lemma.**

(a) $\text{Hom}(M, N_1 \oplus N_2) = \text{Hom}(M, N_1) \oplus \text{Hom}(M, N_2)$.

(b) $\text{Hom}(M_1 \oplus M_2, N) = \text{Hom}(M_1, N) \oplus \text{Hom}(M_2, N)$.

Proof. (a) $\text{Hom}(M, N_1)$, $\text{Hom}(M, N_2)$, as homomorphisms of M into submodules of $N_1 \oplus N_2$, can be regarded as submodules of $\text{Hom}(M, N_1 \oplus N_2)$.

Let δ_i be the decomposition operators: $N \overset{\delta_i}{\twoheadrightarrow} N_i$, $i = 1, 2$. Now for $\sigma \in \text{Hom}(M, N_1 \oplus N_2)$ define $\sigma_1 \in \text{Hom}(M, N_1)$, $\sigma_2 \in \text{Hom}(M, N_2)$ thus:

$$\forall m \in M, \qquad m\sigma_1 = m\sigma\delta_1, \qquad m\sigma_2 = m\sigma\delta_2.$$

This implies

$$m\sigma_1 + m\sigma_2 = m\sigma(\delta_1 + \delta_2) = m\sigma.$$

Conversely, given σ_1, σ_2 we get back σ by using the last relation. Therefore

$$\sigma = \sigma_1 + \sigma_2 \qquad \text{and} \qquad \operatorname{Hom}(M, N_1 \oplus N_2) = \operatorname{Hom}(M, N_1) + \operatorname{Hom}(M, N_2).$$

If $\rho \in \operatorname{Hom}(M, N_1) \cap \operatorname{Hom}(M, N_2)$, then $m\rho \in N_1 \cap N_2 = 0$, $\forall m \in M$. Therefore

$$\rho = 0 \qquad \text{and the sum is direct.}$$

(b) $\forall \sigma \in \operatorname{Hom}(M_1 \oplus M_2, N)$ define σ_1, $\sigma_2 : m_1\sigma_1 = m_1\sigma$, $\forall m_1 \in M_1$, and $M_2\sigma_1 = 0$; similarly $M_1\sigma_2 = 0$, $m_2\sigma_2 = m_2\sigma$, $\forall m_2 \in M_2$. Then

$$\forall\, m = m_1 + m_2 \in M_1 \oplus M_2 : (m_1 + m_2)(\sigma_1 + \sigma_2) = m_1\sigma_1 + m_2\sigma_2 = (m_1 + m_2)\sigma.$$

$$\therefore \quad \sigma_1 + \sigma_2 = \sigma.$$

Moreover, given σ_1, σ_2 we get back σ by using the same relation. Since σ_i can be regarded as an element of $\operatorname{Hom}(M_i, N)$, $i = 1, 2$, this gives

$$\operatorname{Hom}(M_1 \oplus M_2, N) = \operatorname{Hom}(M_1, N) + \operatorname{Hom}(M_2, N)$$

If $\rho \in \operatorname{Hom}(M_1, N) \cap \operatorname{Hom}(M_2, N)$, then $(m_1 + m_2)\rho = 0$, $\forall m \in M_1 \oplus M_2$. Therefore

$$\rho = 0 \qquad \text{and the sum is direct.}$$

(11.5) ***Corollary.***

$$i(M, N_1 \oplus N_2) = i(M, N_1) + i(M, N_2)$$
$$i(M_1 \oplus M_2, N) = i(M_1, N) + i(M_2, N).$$

(11.6) **Lemma.** $\operatorname{Hom}(R, M) \cong M$ *and hence* $i(R, M) = \dim M$.

Proof. $\forall \sigma \in \operatorname{Hom}(R, M)$ define the mapping $\pi : \sigma \to 1\sigma = m_0 \in M$. Now $r\sigma = (1r)\sigma = 1\sigma r = m_0 r$. Conversely, given $m_0 \in M$ we have the mapping $\hat{\sigma} : r\hat{\sigma} = m_0 r$, $\forall r \in R$, so that $\hat{\sigma} \in \operatorname{Hom}(R, M)$ and $\hat{\sigma} = \sigma$. Thus π is a one-one mapping of $\operatorname{Hom}(R, M)$ *onto* M.

But since $(\sigma + \tau)\pi = 1(\sigma + \tau) = 1\sigma + 1\tau = (\sigma)\pi + (\tau)\pi$, π is an isomorphism and

$$\operatorname{Hom}(R, M) \cong M, \qquad i(R, M) = \dim M.$$

(11.7) **Lemma** (Shur's Lemma). $\operatorname{Hom}(M, M)$ *is a ring, and if* M *is irreducible, a skew field over* F.

Proof. Lemma 11.2 showed that $\operatorname{Hom}(M, M)$ is a module over F. If $\sigma, \tau \in \operatorname{Hom}(M, M)$ define $\sigma\tau : m(\sigma\tau) = (m\sigma)\tau$, $\forall m \in M$. Since

$$(m + m')\sigma\tau = ((m + m')\sigma)\tau = (m\sigma + m'\sigma)\tau$$
$$= (m\sigma)\tau + (m'\sigma)\tau = m(\sigma\tau) + m'(\sigma\tau).$$

$$\therefore \quad \sigma\tau \in \operatorname{Hom}(M, M) \qquad \text{and} \qquad \operatorname{Hom}(M, M) \quad \text{is a ring.}$$

Now let M be irreducible, $\sigma \in \operatorname{Hom}(M, M)$. If $\sigma \neq 0$, $M\sigma \neq 0$ and is a submodule of M. Thus $M\sigma = M$, i.e., σ is *onto*. If $M_0 = \text{kernel } \sigma = \{m|, m\sigma = 0\}$, then M_0 is a submodule. Since $M_0 \neq M$, therefore $M_0 = 0$ and σ is an automorphism. Thus $\sigma \neq 0 \Rightarrow \sigma^{-1}$ exists and so $\operatorname{Hom}(M, M)$ is a skew field. Let μ be a representation of R and M its associated representation module. Let $\pi \in \operatorname{Hom}(M, M)$. Then since

$$(mr)\pi = (m\pi)r,$$
$$\therefore \quad m\mu(r)\pi = (m\pi)\mu(r) = m(\pi\mu(r))$$
$$\mu(r)\pi = \pi\mu(r), \qquad \forall r \in R,$$

and this holds $\forall \pi \in \operatorname{Hom}(M, M)$. For this reason we have:

(11.8) **Definition.** $\operatorname{Hom}(M, M)$ *is called the* ***commuting ring*** *of the representation* μ. *We denote it by* $C(\mu)$ *or* $C(M)$.

More generally, if $\pi \in \operatorname{Hom}(M, N)$, $\mu(r)\pi = \pi\nu(r)$ where μ, ν are the representations associated with M and N, respectively. For this reason $\operatorname{Hom}(M, N)$ is sometimes called the *intertwining module* of M and N, and the corresponding representations μ and

ν are said to be intertwined. Recall that a field F is algebraically closed if every equation

$$a_0x^n + a_1x^{n-1} + \cdots + a_{n-1}x + a_n = 0, \qquad a_i \in F,$$

has all its roots in F.

(11.9) **Lemma.** *If μ is an irreducible representation of a ring R over an algebraically closed field F, then rank $C(\mu) = 1$.*

Proof. Let M be the representation module for μ. Since M is irreducible $C(\mu)$ is a skew field by Lemma 11.7. Let $0 \neq \tau \in C(\mu)$. Now $\forall f \in F$, $\forall m \in M$, $\forall r \in R$:

$$\begin{aligned} m(f1 - \tau)\mu(r) &= m(f1 - \tau)r = (fm - m\tau)r = fmr - m\tau r \\ &= fmr - mr\tau = mr(f1 - \tau) = m\mu(r)(f1 - \tau). \end{aligned}$$

$$\therefore \quad (f1 - \tau)\mu(r) = \mu(r)(f1 - \tau), \qquad \forall r \in R.$$

$$\therefore \quad (f1 - \tau) \in C(\mu).$$

Now since F is algebraically closed $\exists f' \in F$, $m' \in M$, $m' \neq 0$ such that

$$m'(f'1 - \tau) = 0,$$

[otherwise $m(1f_1 - \tau)(1f_2 - \tau) \cdots (1f_s - \tau) \neq 0$, $\forall m$, $\forall f_i$, whereas it must $=0$ if the f_i are roots of the polynomial $p(x)$ for which $p(\tau)$ expresses the linear dependence which must subsist between m, $m\tau$, $m\tau^2$, ..., $m\tau^s$, for some s]. Thus, since $m' \neq 0, f'1 - \tau = 0$, therefore

$$\tau = f'1,$$

proving that rank $C(\mu) = 1$. Let R_1 be an indecomposable component of the ring R, $R = R_1 \oplus \cdots$. Let R_1' be the unique maximal right ideal in R_1 [Theorem 7.2(a)]. Suppose that the representation module M of a representation μ has the composition series: $M = M_0 \supset M_1 \supset \cdots \supset M_l = 0$. Then the factors M_{i-1}/M_i are representation modules for the irreducible constituents of μ. We have:

(11.10) ***Theorem.***

$$i(R_1, M) = qk$$

where q is the number of factors of $M \cong R_1/R_1'$ and k is the rank of the skew field $C(R_1/R_1')$.

Proof. Consider the scheme of homomorphic mappings:

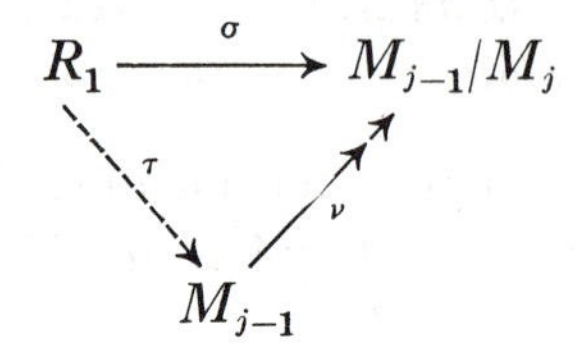

where ν is the natural homomorphism of M_{j-1} onto M_{j-1}/M_j. If τ is any homomorphism of R_1 into M_{j-1}, then $\tau\nu = \sigma$ is a homomorphism of R_1 into M_{j-1}/M_j. Conversely, by Lemma 6.3 $\forall\, \sigma$ there is a τ such that $\tau\nu = \sigma$. Hence the mapping $\Gamma : \tau \to \tau\nu$ is a mapping of $\mathrm{Hom}(R_1, M_{j-1})$ onto $\mathrm{Hom}(R_1, M_{j-1}/M_j)$. Since $(\tau + \tau')\Gamma = (\tau + \tau')\nu = \tau\nu + \tau'\nu = \tau\Gamma + \tau'\Gamma$, Γ is a homomorphism. Its kernel is all τ which map $R_1 \to M_j$ that is all $\tau \in \mathrm{Hom}(R_1, M_j)$. Hence

$$\mathrm{Hom}(R_1, M_{j-1})/\mathrm{Hom}(R_1, M_j) \cong \mathrm{Hom}(R_1, M_{j-1}/M_j).$$

Taking ranks on both sides we have

$$i(R_1, M_{j-1}) - i(R_1, M_j) = i(R_1, M_{j-1}/M_j)$$

and summing over j:

$$(*) \qquad i(R_1, M) = \sum_{j=1}^{l} i(R_1, M_{j-1}/M_j).$$

Now either $i(R_1, M_{j-1}/M_j) = 0$, or $\exists \sigma \in \mathrm{Hom}(R_1, M_{j-1}/M_j)$ such that $R_1\sigma \neq 0$. But then $R_1\sigma = M_{j-1}/M_j$ since the latter is irreducible. If $R_2' =$ Kernel of σ, then $R_1/R_2' \cong M_{j-1}/M_j$, implying that R_2' is maximal in R_1 which in turn implies that $R_2' = R_1'$ by Theorem 7.2(a). Thus σ induces an isomorphism $\hat{\sigma} : R_1/R_1' \cong M_{j-1}/M_j$. Conversely, given $\hat{\sigma}$ we can find

$\sigma : R_1 \to M_{j-1}/M_j$ with kernel R_1', from $\sigma = \nu_1\hat{\sigma}$ where ν_1 is the natural homomorphism $R_1 \twoheadrightarrow R_1/R_1'$. Thus

$$\begin{aligned}\mathrm{Hom}(R_1, M_{j-1}/M_j) &\cong \mathrm{Hom}(R_1/R_1', M_{j-1}/M_j)\\ &\cong \mathrm{Hom}(R_1/R_1', R_1/R_1').\end{aligned}$$

Then $i(R_1, M_{j-1}/M_j) = i(R_1/R_1', R_1/R_1') = k$, the rank of $C(R_1/R_1')$. We have now proved that $i(R_1, M_{j-1}/M_j) = 0$ or k, according as R_1/R_1' is or is not isomorphic to M_{j-1}/M_j. Applied to $(*)$ this result gives the theorem.

12. Multiplicities of the Indecomposable Components in the Regular Representation

The results of the last section can be used to prove:

(12.1) **Theorem.** *Let R be an algebra over a field F. Suppose that*

(12.2) $$R = (R_{11} \oplus \cdots \oplus R_{1q_1}) \oplus \cdots \oplus (R_{s1} \oplus \cdots \oplus R_{sq_s})$$

where the R_{ij} are indecomposable right ideals and the summation (direct) is arranged so that the q_j terms in the jth *block are isomorphic and $j = 1, 2, ..., s$. Thus $R_{jk} \cong R_{im}$ if and only if $j = i$. If*

n = *the rank of R*

d_j = *the rank of R_{j1}*

n_j = *the rank of R_{j1}/R_{j1}', R_{j1}' being the unique maximal right ideal in R_{j1}*

k_j = *the rank of the skew field $C(R_{j1}/R_{j1}')$.*

then

$$n = \sum_{j=1}^{s} (n_j/k_j)d_j .$$

Remark. Theorem 7.2(b) and (c) shows that the R_{j1}/R_{j1}', $j = 1, 2, ..., s$, are representation modules for *all* the distinct irreducible representations. In the language of representations the n_j are the *degrees* of the irreducible representations, the d_j the *degrees* of the principal indecomposable representations and n is the *degree* of the regular representation.

Proof. By Lemma 11.6 $i(R, R_{j1}/R'_{j1}) = \operatorname{rank}(R_{j1}/R'_{j1}) = n_j$. Thus

$$n_j = i(R, R_{j1}/R'_{j1}) = i\left(\sum_{a,b} R_{ab}, R_{j1}/R'_{j1}\right)$$

where the R_{ab} run through the terms of (12.2). Therefore

$$n_j = \sum_{a,b} i(R_{ab}, R_{j1}/R'_{j1})$$

by Corollary 11.5. But $i(R_{ab}, R_{j1}/R'_{j1}) = k_j$ or 0 according as $R_{ab} \cong R_{j1}$ or not. Therefore

$$n_j = \sum_{b=1}^{q_j} i(R_{jb}, R_{j1}/R'_{j1}) = q_j k_j . \tag{12.3}$$

Now counting ranks on both sides of (12.2) we get $n = \Sigma_{j=1}^{s} q_j d_j$, so that

$$n = \sum_{j=1}^{s} (n_j/k_j) d_j , \tag{12.4}$$

proving the theorem.

(12.5) **Corollary.** $n \geqslant \Sigma_{j=1}^{s} n_j^2/k_j$.

Proof. rank $R_{j1}/R'_{j1} \leqslant$ rank R_{j1} or $n_j \leqslant d_j$ and the result follows from (12.4).

(12.6) **Corollary.** $n = \Sigma_{j=1}^{s} n_j^2/k_j$, *if and only if R is semisimple.*

Proof. R is semisimple $\Leftrightarrow$ radical $N = 0 \Leftrightarrow R'_{j1} = 0$ (Theorem 8.4) $\Leftrightarrow d_j = n_j$. Then (12.4) gives the result.

(12.7) **Corollary.** *If R is over an algebraically closed field F, then $n \geqslant \Sigma_{j=1}^{s} n_j^2$ and equality holds if and only if R is semisimple.*

Proof. This follows from Corollaries 12.5 and 12.6 and the fact that $k_j = 1$ by Lemma 11.9.

In the course of proving Theorem 12.1 it was shown that $q_j = n_j/k_j$. We state this result as:

(12.8) **Corollary.** *The number q_j of indecomposable components of the regular representation of a ring R which are isomorphic to a particular indecomposable component R_j is given by $q_j = n_j/k_j$ where $n_j =$ rank R_j/R'_j, $k_j =$ rank of the skew field $C(R_j/R'_j)$, and R'_j is the unique maximal right ideal of R contained in R_j.*

13. The Generalized Burnside Theorem

(13.1) **Theorem.** *Let R be an irreducible algebra of $n_1 \times n_1$ matrices over a field F. Let R have rank n. Then R has an identity and $n = n_1^2/k_1$, where k_1 is the rank of $C(R_1)$, the commuting algebra of R_1.*† *Moreover R is isomorphic to a complete matrix algebra of degree $n/n_1 = n_1/k_1$ over a skew field which itself has rank k_1 over F.*

Proof. If the $n_1 \times n_1$ identity matrix $I \notin R$ let R^* be the algebra generated by I and R. Its elements are finite sums of finite products of elements taken from R or the set of scalar matrices fI, $f \in F$. Clearly R^* is irreducible, otherwise $R \subset R^*$ would be reducible. Also R^* is its own faithful matrix representation. This makes R^* simple (see Lemma 13.2). But now R is a two-sided ideal of R^* and we have a contradiction. Hence $I \in R$ and R as its own faithful irreducible representation is simple:

$$(*) \qquad \therefore \quad R = R_1 \oplus \cdots \oplus R_m,$$

and all R_i are isomorphic right ideals (Corollary 10.5).

$$\therefore \quad i(R, R_1) = i\left(\sum_{j=1}^{m} R_j, R_1\right) = mk_1.$$

$$\therefore \quad n_1 = mk_1,$$

† We may regard the $n_1 \times n_1$ matrices as an irreducible representation provided by R_1.

and, counting ranks on both sides of (*):

$$n = mn_1 ,$$

so that

$$n = n_1^2/k_1$$

as required. (Also follows directly from Corollary 12.6.) Finally from the structure Theorem 10.6 and (*) above it follows that R is isomorphic to a complete matrix algebra of degree m over the skew field $C(R_1)$. Since $m = n/n_1$ and rank $C(R_1) = k_1$ this is the required result.

It remains to prove:

(13.2) **Lemma.** *If a ring R has a faithful irreducible representation μ it is simple.*

Proof. If $n \in N$, the radical of R, then $\mu(n) = 0$ and so $n = 0$. Hence R is semisimple and so by Theorem 10.3:

$$R = A_1 \oplus \cdots \oplus A_s ,$$

where A_i are simple two-sided ideals, each the direct sum of minimal right ideals and, moreover, $A_i A_j = 0$, $i \neq j$. Suppose $R_i \subset A_i$ is the minimal right ideal which affords the representation μ. Then $R_i A_j = 0$, $\forall j \neq i$, and so $\mu(A_j) = 0$, $\forall j \neq i$. But then by the faithfulness of μ, $A_j = 0$, $\forall j \neq i$:

$$R = A_i$$

and R is simple.

EXERCISES

1. Determine the principal components of the ring R

$$R = \left\{ \begin{pmatrix} a & 0 & 0 \\ x & b & 0 \\ z & y & c \end{pmatrix} \right\}$$

of matrices with real coefficients and find the unique maximal right ideal contained in each component. What is the radical of R? [Cf. Example, Section 8.]

2. Let F be the field of real numbers and let

$$R = \left\{ r = \begin{pmatrix} a & -b \\ b & a \end{pmatrix} : a, b \in F \right\}.$$

Verify the Burnside theorem for this case. What is the skew field?

3. Exercise 2 with

$$R = \left\{ \begin{pmatrix} a & b & c & d \\ -b & a & -d & c \\ -c & d & a & -b \\ -d & -c & b & a \end{pmatrix} \right\}.$$

Compare with

$$R = \left\{ \begin{pmatrix} z_1 & z_2 \\ -\bar{z}_2 & \bar{z}_1 \end{pmatrix} \right\}$$

over the complex field.

4. Let R be as in Exercise 2 but let F be the field of complex numbers. Show that now R is not simple but that

$$J = \left\{ \begin{pmatrix} z & iz \\ -iz & z \end{pmatrix} \right\}$$

is an ideal. Is R semisimple?

5. Let

$$R = \left\{ r = \begin{pmatrix} a & 0 \\ b & c \end{pmatrix}, \quad a, b, c \in F \right\}.$$

Is R its own regular representation? Observing that $\rho_1(r) = a$, $\rho_2(r) = c$ are two irreducible representations and that $F = C(\rho_1) = C(\rho_2)$ use the relations (12.5) and (12.6) to prove that R is not semisimple and that there are no other irreducible representations of R.

6. If R_1 is a direct summand of the ring R and M_1 is an admissable submodule of the R-module M prove that the intertwining numbers are related as follows:

$$i(R_1, M) = i(R_1, M_1) + i(R_1, M/M_1).$$

CHAPTER III

The Representation Theory of Finite Groups

14. The Group Algebra

The concept of the representation of a group was discussed in Section 1 of Chapter I. In the present chapter the representation theory of algebras which was developed in Chapter II will be applied to the theory of the representations of finite groups. The connection between the theories is provided by the concept of the group algebra.

(14.1) **Definition.** *The group algebra. Let* $G = \{g_1, g_2, \ldots, g_n\}$ *be a group of finite order n, and let F be an arbitrary field. Denote by* $A(G)$ *an n-dimensional vector space over F in which the elements of a basis are labeled* $g_1, g_2, \ldots, g_n$.

Thus every element $a \in A(G)$ can be written uniquely:

$$a = \sum_{i=1}^{n} f_j g_j, \quad f_i \in F.$$

Now define a multiplication in $A(G)$ as follows: if $a' = \Sigma_{j=1}^{n} f'_j g_j$, then

$$aa' = \sum_{i,j=1}^{n} f_i f'_j g_{ij}, \qquad g_{ij} = \text{the group element } g_i g_j.$$

It is a routine matter to check the validity of the rules:

$$(aa')a'' = a(a'a''); \qquad (a + a')a'' = aa'' + a'a'';$$

$$a''(a + a') = a''a + a''a' \qquad \text{and} \qquad ag_1 = g_1 a,$$

if g_1 is the identity of G. Thus $A(G)$ is an algebra, with an identity, of rank n over F. It is the *group algebra of G over F*.

The relation between representations of G and of $A(G)$ is given by:

(14.2) **Lemma.** *Every representation σ of $A(G)$ induces a representation $\hat{\sigma}$ of G and conversely. The correspondence $\sigma \leftrightarrow \hat{\sigma}$ is unique. Moreover σ is reducible (decomposable) if and only if $\hat{\sigma}$ is reducible (decomposable).*

Proof. Given σ take $\hat{\sigma}$ to be the restriction of σ to the elements of G. Since $\sigma(g_i g_j) = \sigma(g_i)\sigma(g_j)$ we have the same relation for $\hat{\sigma}$ so that $\hat{\sigma}$ is a representation of G. Conversely, if $\hat{\sigma}$ is given, define $\sigma(a) = \Sigma_{i=1}^{n} f_i \hat{\sigma}(g_i)$, when $a = \Sigma_{i=1}^{n} f_i g_i$. Then if $a' = \Sigma_{j=1}^{n} f'_j g_j$, we have

$$\sigma(a)\sigma(a') = \Big(\sum_{i=1}^{n} f_i \hat{\sigma}(g_i)\Big)\Big(\sum_{j=1}^{n} f'_j \hat{\sigma}(g_j)\Big) = \sum_{i,j=1}^{n} f_i f'_j \hat{\sigma}(g_i)\hat{\sigma}(g_j)$$

$$= \sum_{i,j=1}^{n} f_i f'_j \hat{\sigma}(g_i g_j) = \sigma(aa').$$

In the same way $\sigma(a + a') = \sigma(a) + \sigma(a')$. If this σ is now restricted to G we get back $\hat{\sigma}$, so that $\sigma \leftrightarrow \hat{\sigma}$ uniquely. Finally observe that if S is a submodule of an F-$A(G)$ module M, then S is admissable under G if and only if S is admissable under $A(G)$. This remark proves the last statement of the lemma.

As a consequence of Lemma 14.2 all the results with respect to reducibility and decomposability which were proved for the representation of an algebra carry over to the representation of a finite group. In particular we have:

(14.3) **Theorem.** *The irreducible constituents $\mu_1, \mu_2, \ldots, \mu_k$ of a representation μ of a group G are fixed in number and are unique up to order and equivalence. The same statement holds for the indecomposable components.*

Proof. The proof follows from Theorems 4.8 and 4.11.

15. The Regular Representation of a Group

Recall that the regular representation of $A(G)$ is the representation ρ afforded by $A(G)$ as its own representation module. The regular representation $\hat{\rho}$ of G is the restriction of ρ to G. Taking $g_1 = 1, g_2, ..., g_n$ as a basis of $A(G)$ we get

$$g_i\hat{\rho}(g) = g_i g, \qquad i = 1, 2, ..., n.$$

Thus, expressed as a matrix:

$$\hat{\rho}(g) = (\delta_{g_i, g_j g})$$

where the element $\delta_{g_i, g_j g}$ at the intersection of the ith row and jth column is the Kronecker δ:

$$\delta_{a,b} = 0, \qquad a \neq b, \qquad \text{and} \qquad \delta_{a,a} = 1.$$

Now for the character χ^{ρ} (Definition 2.1) of the regular representation of G we have, since $g_i g = g_i$ if and only if $g = 1$,

$$\text{(15.1)} \qquad \chi^{\hat{\rho}}(g) = 0, \qquad g \neq 1; \qquad \chi^{\hat{\rho}}(1) = n.$$

Finally, note that the regular representation is faithful, since

$$\hat{\rho}(g) = 1 \Rightarrow g_i g = g_i, \qquad \forall g_i \Rightarrow g = 1.$$

From Theorem 7.2(c) and (d), valid for $A(G)$, we get directly:

(15.2) **Theorem.** (a) *Every irreducible representation of a group G occurs as an irreducible constituent of the regular representation.*

(b) *The number of irreducible representations of G is the same as the number of principal indecomposable representations.*

16. Semisimplicity of the Group Algebra

(16.1) **Theorem.** *Let G be a finite group of order n. Let F be a field of characteristic p. The group algebra $A(G)$ of G over F is semisimple if and only if p does not divide n, or $p = 0$.*

Proof. (a) $p = 0$ or $p \nmid n$. Let N be the radical of $A(G)$ and suppose

$$r = \sum_{j=1}^{n} f_j g_j, \qquad r \in N. \tag{16.2}$$

If ρ is the regular representation of $A(G)$ we get

$$\rho(r) = \sum_{j=1}^{n} f_j \rho(g_j) = \sum_{j=1}^{n} f_j \hat{\rho}(g_j),$$

$\hat{\rho}$ the regular representation of G. Taking the traces of the matrices on both sides, (15.1) gives

$$\chi^{\rho}(r) = f_1 \chi^{\hat{\rho}}(g_1) = f_1 n. \tag{16.3}$$

However Lemma 4.14 showed that

$$\chi^{\rho}(r) = \chi^{\rho_i}(r) + \cdots + \chi^{\rho_k}(r),$$

ρ_i all irreducible constituents. But since $r \in N$, $\rho_i(r) = 0$ and $\chi^{\rho_i}(r) = 0$; hence $\chi^{\rho}(r) = 0$ and (16.3) becomes

$$0 = f_1 n. \tag{16.4}$$

Since $p = 0$ or $p \nmid n$ we have $f_1 = 0$. Now N is an ideal and $rg_i^{-1} \in N$. Again,

$$rg_i^{-1} = \left(\sum_{j=1}^{n} f_j g_j\right) g_i^{-1} = f_i g_1 + \sum_{j \neq 1} f_j g_j g_i^{-1},$$

so that by the same argument $f_i = 0$, $\forall i$. Therefore $r = 0$ and hence $N = 0$, showing that $A(G)$ is semisimple.

(b) p/n. Let $r = \Sigma_{j=1}^{n} g_j$. Now $\forall i$, $rg_i = \Sigma_{j=1}^{n} g_j g_i = r$, since the terms of the summation are again the groups elements in some order. Similarly $g_i r = r$. Thus Fr is a two-sided ideal. Now

$$r^2 = \sum_{i=1}^{n} \left(\sum_{j=1}^{n} g_j\right) g_i = r + r + \cdots + r = nr = 0.$$

Therefore Fr is a nilpotent two-sided ideal. By Lemma 8.5: $N \supset Fr$ so that $N \neq 0$ and $A(G)$ is nonsemisimple.

(16.5) **Corollary.** *A representation μ of a finite group G of order n is completely reducible over a field of characteristic p if $p = 0$ or $p \nmid n$.*

Proof. The proof follows from Theorem 9.3, since $A(G)$ is semisimple.

(16.6) **Corollary.** *If G is of order n and $n_1, n_2, ..., n_s$ are the degrees of the distinct irreducible representations $\rho_1, \rho_2, ..., \rho_s$ of G over an algebraically closed field F of characteristic p, $p = 0$, or $p \nmid n$, then*

$$n = n_1^2 + n_2^2 + \cdots + n_s^2 . \tag{16.7}$$

Proof. The proof follows from Corollary 12.7, since $A(G)$ is semisimple.

Remark. Every group G has the trivial irreducible representation $\rho_1(g) = 1$, $\forall g \in G$. Thus in (16.7) we always have $n_1 = 1$.

(16.8) **Corollary.** *In the regular representation ρ of a finite group G of order n over an algebraically closed field F of characteristic p, $p = 0$, or $p \nmid n$, each irreducible representation ρ_j occurs as often as its degree n_j.*

Proof. $A(G)$ is semisimple by Theorem 16.1 so that the indecomposable components of $A(G)$ are irreducible. Let R_j be the right ideal which affords the irreducible representation ρ_j, then according to Corollary 12.8 the number of irreducible constituents of ρ which are equivalent to ρ_j is

$$q_j = n_j/k_j, \qquad n_j = \text{rank of } R_j = \text{degree of } \rho_j$$

$$k_j = \text{rank of the skew field } C(R_j).$$

Since F is algebraically closed, $k_j = 1$ by Lemma 11.9. This proves the result. An immediate consequence is:

(16.9) **Corollary.** *Under the conditions of Corollary* 16.8 *we have*

$$\chi^{\rho}(r) = \sum_{j=1}^{s} n_j \chi_j^{\rho}(r), \qquad \forall r \in A(G), \tag{16.10}$$

where χ^{σ} denotes the character of the representation σ.

Proof. Apply Lemma 4.14 and the fact that equivalent representations have the same character.

Remark. Let F be a subfield of a field E. Then E is called an extension field of F. Let M be an E-R module for the ring R. The example on p. 64 shows that M regarded as an F-R module may be irreducible while being reducible as an E-R module. In such a case we say that the representation *splits* on extension of the field.

(16.11) **Definition.** *A representation μ of a ring R in a field F is absolutely irreducible if it remains irreducible (does not split) in any extension E of the ground field F.*

Now let G be a finite group of order n and let F be an algebraically closed field of characteristic p, $p = 0$, or $p \nmid n$. If $F \subset E$ we know from the theory of fields that the field E can be extended to an algebraically closed field E^*, and all three fields necessarily have the same characteristic p. Thus $F \subset E \subset E^*$ and any representation σ in F can be considered as a representation in E^*. Moreover, because of the condition on p, $A(G)$ is semisimple over each field. By Corollary 16.8 the irreducible representation ρ_j of $A(G)$ in F occurs n_j times (as often as its degree) in the regular representation ρ of $A(G)$ in F. Now if ρ_j is reducible in E^* and γ_j is an irreducible constituent of ρ_j, then γ_j occurs at least n_j times. But this is more than its degree (since degree $\gamma_j <$ degree ρ_j). However, E^* itself is algebraically closed and we have a contradiction of Corollary 16.8. Consequently ρ_j is also irreducible over E^* and hence in E. We have:

(16.12) **Lemma.** *An irreducible representation of the group algebra of a finite group G of order n in an algebraically closed field F of characteristic p, $p = 0$, or $p \nmid n$ is absolutely irreducible.*

Throughout the remainder of the chapter, F and G will satisfy the conditions of Corollary 16.8. Thus all the irreducible representations of $A(G)$ will be absolutely irreducible.

Example. Let G be a cyclic group of order 4: $G = \langle g : g^4 = 1 \rangle$. Let F be the field of rational numbers and let E be the field of complex numbers, $x + iy$, having x and y rational. For the ring R take $A(G)$ over F. Then R consists of linear combinations of the elements of G with coefficients in F. The E-R module M is the two-dimensional vector space over E with basis $e_1 = (1, 0)$, $e_2 = (0, 1)$. To give the effect of $A(G)$ on M it suffices to give the effect of g on the basis of M. Let

$$e_1 g = e_1 + 2e_2 , \qquad e_2 g = -e_1 - e_2 .$$

We assert:

(1) M is irreducible as an F-R module.

Proof. Let $m = f_1 e_1 + f_2 e_2$, $f_i \in F$, m any vector in M. Then

$$m(\lambda g + \mu g^2) = [\lambda(f_1 - f_2) - \mu f_1]e_1 + [\lambda(2f_1 - f_2) - \mu f_2]e_2$$

and λ, μ may be chosen so that this equals $fe_1 + f'e_2$ for arbitrarily prescribed f, f'. Thus $mR = M$ and M has no proper submodule.

(2) M is reducible as an E-R module.

Proof. Let $m = e_1 + (1 - i)e_2$. It is easily verified that $mg = im$. Hence Em is a proper R-submodule of M and M is reducible.

17. The Center of the Group Algebra

(17.1) **Definition.** *The set of all elements z of the group algebra $A(G)$ which commute with each element of $A(G)$ is called the Center of the Group Algebra. If Z denote the center of $A(G)$ we have*

$$Z = \{z : xz = zx, \quad \forall x \in A(G)\}.$$

Since $z, z' \in Z$, $f \in F \Rightarrow zz', z + z', fz \in Z$, therefore Z is a subalgebra of $A(G)$.

Now let $z \in Z$:

$$(17.2)\quad z = \sum_{i=1}^{n} f_i g_i , \qquad f_i \in F, \qquad g_i \in G, \qquad G:1 = n.$$

$\forall g \in G$ we have

$$gzg^{-1} = z = \sum_{i=1}^{n} f_j g_j = \sum_{j=1}^{n} f_j g g_j g^{-1},$$

and equating coefficients we see that

$$f_i = f_j \qquad \text{if} \qquad g_i = gg_jg^{-1}.$$

Thus all group elements which are conjugate have the same coefficient in (17.2). We may write

$$(17.3)\qquad z = f_1C^1 + f_2C^2 + \cdots + f_sC^s$$

where C^i denotes the sum of all the elements in a certain class of conjugate group elements. Conversely, any element z given by (17.3) belongs to Z since

$$gC^ig^{-1} = C^i,$$

the left-hand side merely rearranging the terms in the sum. This gives:

(17.4) **Lemma.** *Z is a subalgebra of rank s where s is the number of classes of conjugate elements of G. It has as a basis $C^1, C^2, \ldots, C^s$; each C^i is the sum of the elements in a class of conjugate elements.*

18. The Number of Inequivalent Irreducible Representations

(18.1) **Theorem.** *The number of inequivalent irreducible representations of the group G in an algebraically closed field F of characteristic p, $p = 0$ or $p \nmid n = G:1$, is equal to the number s of classes of conjugate elements in G.*

Proof. Let $\rho_1, \rho_2, ..., \rho_k$ be the k distinct irreducible representation of $A(G)$. Let $z \in Z$. Since $xz = zx$:

$$\rho_i(x)\rho_i(z) = \rho_i(z)\rho_i(x), \qquad \forall\, x \in A(G).$$

Thus $\rho_i(z) \in C(\rho_i)$, the commuting algebra of ρ_i (Definition 11.8). Since ρ_i is irreducible and F algebraically closed, $C(\rho_i)$ is a skew field of rank 1 over F (Lemma 11.9). Thus $\forall\, \rho_i$, $i = 1, 2, ..., k$,

$$\rho_i(z) = f_i 1, \qquad f_i \in F. \tag{18.2}$$

Then $\forall\, z \in Z$ we have a mapping $\Gamma : z \to (f_1, f_2, ..., f_k)$. Since $\rho_i(fz+f'z') = f\rho_i(z)+f'\rho_i(z') = ff_i 1 + f'f_i' 1 = (ff_i + f'f_i')1$, Γ is a homomorphism of Z as a module into V_k, a k-dimensional vector space over F. If $z \to (0, 0, ..., 0)$, then (18.2) gives $\rho_i(z) = 0$, $\forall\, i$ so that $z \in N$, the radical of $A(G)$. Since $A(G)$ is semisimple, $N = 0$ and hence $z = 0$. Thus Γ is an *isomorphism*.

Finally, let $(f_1, f_2, ..., f_k)$ be an arbitrary vector of V_k. If 1 is the identity of $A(G)$, then $z_i = f_i 1 \in Z$ and $\rho_i(z_i) = f_i 1$. However, by the Frobenius-Shur theorem (10.10) $\exists z$ such that

$$\rho_i(z) = \rho_i(z_i) = f_i 1, \qquad \forall i,$$

and hence

$$z \to (f_1, f_2, ..., f_k).$$

Thus Γ is an isomorphism of Z *onto* V_k, and so dim V_k = rank Z $= s$, the number of classes of conjugate elements of G (Lemma 17.4). Therefore

$$k = s$$

and the theorem is proved.

(18.3) **Corollary.** *The irreducible representations of an abelian group are all of degree one. Moreover, their number equals the order of the group.*

Proof. By (16.7),

$$n = n_1^2 + \cdots + n_s^2 .$$

Since the group is abelian, each element is its only conjugate and $s = n$. Therefore

$$n_i = 1.$$

19. Relations on the Irreducible Characters

Notation. χ^i is the character of the irreducible representation ρ_i of degree n_i, $i = 1, 2, ..., s$:

$C_1, C_2, ..., C_s$ denote the classes of conjugate elements of G

C_1 is the class consisting of only the identity $1 \in G$.

C^i is the sum of the elements of the class C_i.

$\chi^i_j = \chi^i(\hat{g}_j)$, $\hat{g}_j$ being a fixed representative of the class C_j (χ^i is constant on C_j).

h_i is the number of elements in the class C_i.

Since $C^i \in Z$, the center of $A(G)$, we have as in (18.2):

(19.1) $\quad \rho_t(C^i) = f^i_t 1, \qquad f^i_t \in F, \qquad 1$ is *the* basis of $C(\rho_t)$ over F.

Thinking in terms of the matrix representations and taking traces on both sides, (19.1) gives

(19.2)
$$\chi^t(C^i) = f^i_t n_t,$$

$n_t =$ the degree of ρ_t, a positive integer. But

$$\chi^t(C^i) = \chi^t\left(\sum_{g \in C_i} g\right) = h_i \chi^t_i,$$

so that

(19.3)
$$f^i_t = (h_i/n_t)\chi^t_i.$$

Now since the C^j are a basis of Z (Lemma 17.4)

(19.4)
$$C^i C^j = \sum_{k=1}^{s} f^{ij}_k C^k, \qquad f^{ij}_k \in F.$$

Actually $f_k^{ij} = l_k^{ij} \cdot 1$, 1 the identity of F and l_k^{ij} = a nonnegative integer. This is because

$$C^iC^j = \sum_{g_i \in C_i} g_i \cdot \sum_{g_j \in C_j} g_j = \sum_{g_i \in C_i, g_j \in C_j} g_i g_j$$

and when the group elements which are the terms of the final sum are collected to form the classes, only integer coefficients can occur. Applying ρ_t to (19.4) we get

$$\rho_t(C^iC^j) = \rho_t(C^i) \cdot \rho_t(C^j) = \sum_{k=1}^{s} f_k^{ij} \rho_t(C^k)$$

and using (19.1)

$$f_t^i \cdot f_t^j \cdot 1 = \left(\sum_{k=1}^{s} f_k^{ij} f_t^k \right) 1 = \left(\sum_{k=1}^{s} l_k^{ij} f_t^k \right) 1$$

so that

$$f_t^i f_t^j = \sum_{k=1}^{s} l_k^{ij} f_t^k . \tag{19.5}$$

Now substitute from (19.3) to get

$$(h_i h_j / n_t^2) \chi_i^t \chi_j^t = \sum_{k=1}^{s} l_k^{ij} (h_k / n_t) \chi_k^t . \tag{19.6}$$

If we multiply (19.6) by n_t^2 and sum over $t = 1, 2, \ldots, s$ there results

$$h_i h_j \sum_{t=1}^{s} \chi_i^t \chi_j^t = \sum_{k=1}^{s} l_k^{ij} h_k \left(\sum_{t=1}^{s} n_t \chi^t(\hat{g}_k) \right). \tag{19.7}$$

However, we saw (16.10) that $\Sigma_{t=1}^{s} n_t \chi^t(\hat{g}_k) = \chi^\rho(\hat{g}_k)$, χ^ρ being the character of the regular representation ρ. But then by (15.1) $\chi^\rho(\hat{g}_k) = 0$ unless $\hat{g}_k = \hat{g}_1 = 1 \in G$ in which case $\chi^\rho(\hat{g}_1) = n$, the order of G. Thus, since $h_1 = 1$ we get from (19.7)

$$h_i h_j \sum_{t=1}^{s} \chi_i^t \chi_j^t = l_1^{ij} n. \tag{19.8}$$

It remains to compute l_1^{ij}. This is the number of times the identity of G occurs in the product C^iC^j. Since the inverses of the elements of a class form a class, called the inverse class,

$$\begin{aligned} l_1^{ij} &= 0, && \text{if } C_i \text{ and } C_j \text{ are not inverse classes} \\ &= h_i = h_j\,, && \text{if } C_i \text{ and } C_j \text{ are inverse classes.} \end{aligned}$$

Introduce the symbol $\delta_{ij^{-1}} = 1, 0$ according as C_i and C_j are, or are not, inverse classes. Then we can write for (19.8)

$$\sum_{t=1}^{s} \chi_i^t \chi_j^t = (n/h_i)\delta_{ij^{-1}}\,. \tag{19.9}$$

This is the *first orthogonal relation* on the irreducible characters of a group.

Consider the $s \times s$ matrix $\psi = (\chi_j^i)$ formed by the values of the s characters χ^i on the s classes C_j:

$$\psi = \begin{pmatrix} \chi_1^1 & \chi_2^1 & \cdots & \chi_s^1 \\ \vdots & \vdots & & \vdots \\ \chi_1^s & \chi_2^s & \cdots & \chi_s^s \end{pmatrix}. \tag{19.10}$$

Let ψ' be the transpose matrix of ψ. Using (19.9) we calculate

$$P = \psi'\psi = (p_{ij});$$

p_{ij}, the element at the intersection of the ith row and jth column, will be $(n/h_i)\delta_{ij^{-1}}$.

Now put $\tilde{p}_{ab} = (h_a/n)\delta_{ab^{-1}}$ and form the matrix $\tilde{P} = (\tilde{p}_{ab})$. Now

$$\sum_{j=1}^{s} p_{ij}\tilde{p}_{jk} = \sum_{j=1}^{s} (n/h_i)\delta_{ij^{-1}}(h_j/n)\delta_{jk^{-1}} = \delta_{ik}$$

because all summands are zero unless C_j is the inverse class of both C_i and C_k; that is, unless $i = k$, in which case there is only the term 1. Thus $P\tilde{P} = I$ so that $\tilde{P} = P^{-1}$. Then $P^{-1} = \psi^{-1}\psi'^{-1}$

or $\psi P^{-1}\psi' = I$. Equating the elements at the intersection of the ith row and jth column on both sides we have

$$\sum_{a,b=1}^{s} \chi_a^i \tilde{p}_{ab} \chi_b^j = \delta_{ij}$$

or

$$(1/n) \sum_{a,b=1}^{s} \chi_a^i h_a \delta_{ab^{-1}} \chi_b^j = (1/n) \sum_{a=1}^{s} h_a \chi_a^i \chi_{a^{-1}}^j = \delta_{ij} .$$

If we use $\hat{g}_a$, the fixed representative of the class C_a we can write

$$(19.11) \qquad (1/n) \sum_{a=1}^{s} h_a \chi^i(\hat{g}_a) \chi^j(\hat{g}_a^{-1}) = \delta_{ij} .$$

Since there are h_a elements in the class C_a and since χ^i is the same on each, (19.11) can be written

$$(19.12) \qquad (1/n) \sum_{g \in G} \chi^i(g) \chi^j(g^{-1}) = \delta_{ij} .$$

This is the *second orthogonal relation* on the irreducible characters.

By means if (19.12) we prove:

(19.13) **Lemma.** *Two representations σ and τ of a group G are equivalent if and only if the corresponding characters are equal.*

Proof. If σ and τ are equivalent we have already seen (Section 2) that $\chi^\sigma = \chi^\tau$. Conversely, assume that these characters are equal. If each irreducible character ρ_i is contained in σ with multiplicity a_i and in τ with multiplicity b_i, then

$$\chi^\sigma = \sum_{j=1}^{s} a_j \chi^j, \qquad \chi^\tau = \sum_{j=1}^{s} b_j \chi^j.$$

But then

$$(1/n) \sum_{g \in G} \chi^\sigma(g) \chi^k(g^{-1}) = \sum_{j=1}^{s} (1/n) \sum_{g \in G} a_j \chi^j(g) \chi^k(g^{-1})$$

$$= \sum_{j=1}^{s} a_j \delta_{jk} = a_k = (1/n) \sum_{g \in G} \chi^\tau(g) \chi^k(g^{-1}) = b_k .$$

Thus σ and τ contain the same irreducible representations with the same multiplicity and so, because of their complete reducibility, they are equivalent.

20. The Module of Characters over the Integers

If M and N are two F-R modules we can form their direct sum $M \oplus N$. This is the set S of pairs (m, n), $m \in M$, $n \in N$, made into an F-R module by the rules

$$\begin{aligned} (m, n) + (m', n') &= (m + m', n + n') \\ f(m, n) &= (fm, fn), \qquad f \in F, \\ (m, n)r &= (mr, nr), \qquad r \in R. \end{aligned}$$

S contains the F-R submodules $M' = \{(m, 0)\}$, $N' = \{(0, n)\}$ and it is clear that $M \cong M'$, $N \cong N'$.

A basis for S is: $(m_1, 0)$, $(m_2\ 0)$, ..., $(m_a, 0)$, $(0, n_1)$, ..., $(0, n_b)$ where the first a elements are a basis of M' and the last b a basis of N'. As a representation module S provides a representation σ and if μ and ν are the representations afforded by M and N we write $\sigma = \mu \oplus \nu$. If $\check{M}(r)$ and $\check{N}(r)$ are the matrices for $\mu(r)$ and $\nu(r)$,

$$\sigma(r) \to \begin{pmatrix} \check{M}(r) & 0 \\ 0 & \check{N}(r) \end{pmatrix} \qquad \text{and so} \qquad \chi^\sigma = \chi^\mu + \chi^\nu.$$

This shows that any linear combination of characters with nonnegative integer coefficients is the character of some representation.

Now let $\mathfrak{A} = \{\Sigma_{j=1}^{s} a_j \chi^j : a_j$ arbitrary rational integers$\}$. The χ^j, $j = 1, ..., s$ are the irreducible characters of the group G.

(20.1) **Definition.** *$\mathfrak{A}$ is the module of generalized characters over the integers with a basis of irreducible characters.*

Since every character is the sum of irreducible characters (4.14) we see that $\mathfrak{A}$ contains all characters. Let us define on $\mathfrak{A}$ a symmetric bilinear form (χ, ψ).

(20.2) **Definition.** *If* $\chi = \Sigma\, a_i \chi^i$, $\psi = \Sigma\, b_j \chi^j$:

(20.3) $(\chi, \psi) = (1/n) \sum_{g \in G} \chi(g)\psi(g^{-1})$, *n the order of G.*

Because of (19.12)

$$(\chi, \psi) = (1/n) \sum_{i,j=1}^{s} \sum_{g \in G} a_i b_j \chi^i(g)\chi^j(g^{-1}) = \sum_{i,j=1}^{s} a_i b_j \delta_{ij} = \sum_{j=1}^{s} a_j b_j .$$

Note that $(\chi^i, \chi^j) = \delta_{ij}$ so that the irreducible characters are an orthonormal basis of $\mathfrak{A}$. Since $(\chi, \chi) = \Sigma_{j=1}^{s} a_j^2$, $(\chi, \chi) = 1$ if and only if $\exists i$ such that $a_i^2 = 1$ and $a_j = 0$ for $j \neq i$. But then either $\chi = \chi^i$ or $\chi = -\chi^i$. This gives the following criterion:

$\chi \in \mathfrak{A}$ is an irreducible character if and only if $(\chi, \chi) = 1$ and

$$\chi(1) > 0.$$

Hence, given a number of characters, we can form arbitrary integral combinations and check for irreducible characters.

Example. The symmetric group S_3 is generated by a, b with the defining relations $a^2 = b^3 = (ab)^2 = 1$. We can see (final example of Section 1) that a character χ is $\chi(1) = 3$, $\chi(a) = \chi(ab) = \chi(ba) = 1$, $\chi(b) = \chi(b^2) = 0$. Now χ^1 being the one-character, $\chi^1(g) = 1$ $\forall g \in G$. We can form $\psi = m\chi + n\chi'$ and impose the conditions $(\psi, \psi) = 1$, $\psi(1) > 0$ to find that $\psi = \chi - \chi^1$ is an irreducible character of degree 2.

Remark. Note that if χ^α is an irreducible character and $\psi = \Sigma_\alpha\, a_\alpha \chi^\alpha$ is any character, then $(\psi, \chi^\alpha) = a_\alpha$ is the multiplicity with which χ^α occurs in ψ. Moreover two irreducible characters χ^α and χ^β are equal if and only if $(\chi^\alpha, \chi^\beta) = 1$.

In the next section a multiplication of characters is defined which makes the module $\mathfrak{A}$ into a *ring of characters over the integers.*

21. The Kronecker Product of Two Representations

Let μ and ν be two representations of a ring R afforded by the F-R modules M and N, respectively. We are going to construct an

F-R module $M \times N$ which will be the representation module for a representation denoted by $\mu \times \nu$ and called the Kronecker product of μ by ν. Let $m_1, \ldots, m_s$ and $n_1, \ldots, n_t$, be a basis of M and N, respectively. Denote by $M \times N$ an arbitrary F-module with a basis of st elements labeled (m_i, n_j), $i = 1, 2, \ldots, s$, $j = 1, 2, \ldots, t$. Suppose that for $r \in R$, $m_i r = \sum_{l=1}^{s} f_i^l m_l$ and $n_j r = \sum_{k=1}^{t} \hat{f}_j^k n_k$. Then $M \times N$ can be made into an F-R module by defining the effect of r on each basis element (m_i, n_j) as follows:

$$(m_i, n_j)r = \sum_{l=1}^{s} \sum_{k=1}^{t} f_i^l \hat{f}_j^k (m_l, n_k).$$

The effect of r is then extended linearly to a general element $u \in M \times N$. The relations

$$(1) \qquad (f\mathring{u})r = f(\mathring{u}r), \qquad f \in F,$$

$$(2) \qquad \mathring{u}(rr') = (\mathring{u}r)r', \qquad r, r' \in R,$$

are readily checked. The relation (2) means, for instance,

$$[\mu \times \nu](rr') = [\mu \times \nu](r) \cdot [\mu \times \nu](r')$$

so that $\mu \times \nu$ is a representation of R.

Let us arrange the basis elements $\{(m_i, n_j)\}$ in t blocks of s elements each:

$$(m_1, n_1), (m_2, n_1), \ldots, (m_s, n_1); \ldots; (m_1, n_j), (m_2, n_j), \ldots, (m_s, n_j);$$
$$\ldots; (m_1, n_t), (m_2, n_t), \ldots, (m_s, n_t).$$

Using this ordered basis we wish to find the matrix for $[\mu \times \nu](r)$. Now

$$(m_\alpha, n_j)[\mu \times \nu](r) = (m_\alpha, n_j)r = (m_\alpha r, n_j r) = \left(\sum_{\beta=1}^{s} f_\alpha^\beta m_\beta, \sum_{k=1}^{t} g_j^k n_k\right)$$

$$= \sum_{\beta=1}^{s} \sum_{k=1}^{t} f_\alpha^\beta g_j^k (m_\beta, n_k).$$

Viewing the $st \times st$ matrix $[\mu \times \nu](r)$ as a $t \times t$ matrix of blocks of $s \times s$ matrices, the relation above shows that we have the element $f_\alpha^\beta g_j^k \in F$ at the intersection of the αth row and βth column of the block occupying the jth row and the kth column of blocks. Keeping j and k fixed we have the matrix $(f_\alpha^\beta)g_j^k$ as the j, kth block matrix.

Since $\mu(r) = (f_\alpha^\beta)$ and $\nu(r) = (g_j^k)$, the rule for writing down the matrix $[\mu \times \nu](r)$ can be stated: replace each entry g_j^k in $\nu(r)$ by the *matrix* $\mu(r)$ with each of its elements multiplied by g_j^k.

For the character $\chi^{\mu\times\nu}$ of $\mu \times \nu$ we get

$$\chi^{\mu\times\nu}(r) = \sum_{j=1}^{t}\left(\sum_{\alpha=1}^{s} f_\alpha^\alpha\right) g_j^j = \text{trace } \mu(r) \sum_{j=1}^{t} g_j^j = \text{trace } \mu(r) \text{ trace } \nu(r)$$
$$= \chi^\mu(r)\chi^\nu(r).$$

This shows that the product of two characters is again a character. Since every character is the sum of irreducible characters,

$$\chi^{\mu\times\nu} = \sum_\alpha c_\alpha^{\mu\nu}\chi^\alpha, \tag{21.1}$$

where $c_\alpha^{\mu\nu} \geqslant 0$ are rational integers. It is clear that $M \times N$ and $N \times M$ are operator isomorphic under the correspondence $(m_i, n_j) \leftrightarrow (n_j, m_i)$.

Similarly $(M \times N) \times K \simeq M \times (N \times K)$, under $((m_i, n_j), k_l) \leftrightarrow (m_i, (n_j, k_l))$. Hence $\mu \times \nu \sim \nu \times \mu$ and $(\mu \times \nu) \times \kappa \sim \mu \times (\nu \times \kappa)$. The same results follow, by Lemma 19.13, from the equality of the characters:

$$\chi^{\mu\times\nu} = \chi^\mu\chi^\nu = \chi^\nu\chi^\mu = \chi^{\nu\times\mu},$$
$$\chi^{(\mu\times\nu)\times\kappa} = \chi^{\mu\times\nu}\chi^\kappa = \chi^\mu\chi^\nu\chi^\kappa = \chi^{\mu\times(\nu\times\kappa)}.$$

Finally it is easy to verify the following rules:

$$\mu \times (\nu_1 \oplus \nu_2) \sim \mu \times \nu_1 \oplus \mu \times \nu_2,$$
$$(\mu_1 \oplus \mu_2) \times \nu \sim \mu_1 \times \nu \oplus \mu_2 \times \nu,$$

where $\oplus$ denotes the direct sum as usual.

EXERCISES

1. If χ is a character show that $\bar{\chi}$, the conjugate complex of χ, is also a character.

2. Let χ^α and χ^β be irreducible characters of a group. Show that $\chi^\alpha\chi^\beta$ does not contain the one-character, unless $\chi^\alpha = \bar{\chi}^\beta$ in which case it contains it exactly once.

3. Can a nonabelian group have all its irreducible characters of degree one?

4. If χ^α, χ^β, χ^γ are irreducible characters of a group show that $\chi^\alpha\bar{\chi}^\beta$ contains χ^γ with the same multiplicity with which χ^β occurs in $\chi^\alpha\bar{\chi}^\gamma$.

Remark. Let σ be a representation of a group over the complex field and let χ be its character. We will frequently need the result

$$\chi(g^{-1}) = \bar{\chi}(g). \tag{21.2}$$

Since $\chi(g)$ is the sum of the characteristic roots of the matrix $\sigma(g)$ the result follows directly from matrix theory. On the other hand, it can be deduced from our results as follows: if $\overline{\sigma(g)}$ denotes the matrix obtained by taking the conjugate complex of the entries of the matrix $\sigma(g)$, it is clear that $\bar{\sigma}(g) = \overline{\sigma(g)}$ provides a representation with character $\bar{\chi}$. Moreover $\bar{\chi}$ is irreducible if and only if χ is irreducible. Again $\tilde{\sigma}(g) = ((\sigma(g))^{-1})'$ (that is, taking the inverse transpose of the matrix $\sigma(g)$) provides a representation with character $\tilde{\chi}$ such that $\tilde{\chi}(g) = \chi(g^{-1})$, and $\tilde{\chi}$ is irreducible if and only if χ is irreducible. Now let χ be irreducible and consider $(\bar{\chi}, \tilde{\chi})$, which is zero unless $\bar{\chi} = \tilde{\chi}$. But

$$(\bar{\chi}, \tilde{\chi}) = (1/n) \sum_{g \in G} \bar{\chi}(g)\tilde{\chi}(g^{-1}) = (1/n) \sum_{g \in G} \bar{\chi}(g)\chi(g) > 0$$

so that $\bar{\chi} = \tilde{\chi}$. Hence $\bar{\chi}(g) = \tilde{\chi}(g) = \chi(g^{-1})$ and our result is proved for irreducible characters. Since every character is the sum of irreducible characters the general case follows easily.

22. Linear Characters

The degree of a character is defined as the degree of the corresponding representation. Characters of degree one are called linear. Since a representation of degree one is a number it coincides with its character. Thus a linear character is a representation whose values, as field elements, commute: $\chi(ab) = \chi(a)\chi(b) = \chi(b)\chi(a)$. How many linear characters does a finite group G have? If G is abelian all its representations are of degree one and so all its characters are linear. Moreover, by Corollary 18.3 their number equals the order of the group. In the general case the answer depends on the *commutator subgroup* of G.

(22.1) **Definition.** *An element $a^{-1}b^{-1}ab$, where a and b are any elements of G is called a commutator. The subgroup G' generated by all the commutators of G is called the commutator subgroup.*

Since $(a^{-1}b^{-1}ab)^{-1} = b^{-1}a^{-1}ba$, the inverse of a commutator is a commutator. Also if $t \in G$,

$$t^{-1}(a^{-1}b^{-1}ab)t = (t^{-1}a^{-1}t)(t^{-1}b^{-1}t)(t^{-1}at)(t^{-1}bt) = x^{-1}y^{-1}xy,$$

so that the transform of a commutator is a commutator. This shows that G' is a normal subgroup of G. Finally G/G' is abelian, since $a^{-1}b^{-1}ab \in G'$ means $ab = ba(a^{-1}b^{-1}ab) \equiv ba \bmod G'$. Note that if G is abelian $G' = 1$. We can now state:

(22.2) **Lemma.** *If G is a group and G' its commutator subgroup then the linear characters of G and G/G' are in one-one correspondence, and the corresponding characters have the same set of values. The number of linear characters of G/G', then also of G, equals $G : G'$, the index of G' in G.*

Proof. Let $\hat{\chi}$ be a linear character of G/G'. Define χ on G as

$$(*) \qquad \chi(g) = \hat{\chi}\lfloor g \rfloor, \qquad \lfloor g \rfloor \text{ the class of } g \text{ modulo } G'.$$

Since

$$\chi(g_1g_2) = \hat{\chi}(\lfloor g_1g_2 \rfloor) = \hat{\chi}(\lfloor g_1 \rfloor \lfloor g_2 \rfloor) = \hat{\chi}(\lfloor g_1 \rfloor)\hat{\chi}(\lfloor g_2 \rfloor) = \chi(g_1)\chi(g_2),$$

therefore χ is a linear character on G. Conversely if χ is a linear character on G, define $\hat{\chi}$ on G/G' by $(*)$. To see that $\hat{\chi}$ is well defined we need only check that if $g \in G'$, then $\chi(g) = 1$, and for this it suffices to have g a commutator, $g = a^{-1}b^{-1}ab$. But $\chi(a^{-1}b^{-1}ab) = (\chi(a))^{-1}(\chi(b))^{-1}\chi(a)\chi(b) = 1$. Thus $\chi, \hat{\chi}$ are mutually defined by $(*)$ and we have that the linear characters of G and G/G' are in one-one correspondence and that corresponding characters have the same set of values. However, G/G' is abelian so that (Corollary 18.3) all its characters are linear and their number equals the order of $G/G' = G : G'$.

EXERCISES

1. Show that the symmetric group S_n, consisting of all permutations on n symbols, has exactly two linear characters.

2. Let F be a finite field of characteristic p having $q = p^n$ elements. The set of all nonsingular 2×2 matrices over F form a group denoted by $GL(2, q)$. Show that this group has $q - 1$ linear characters over the complex field. [*Note*: the order of $GL(2, q)$ is $q(q - 1)(q^2 - 1)$. See Dickson [11], p. 77.]

23. Induced Representations and Induced Characters

Let G be a group and H a subgroup of G. If σ is a representation of G then σ restricted to H clearly gives a representation of H. We consider now the opposite question: given a representation σ of H can a representation σ^* of G be constructed? For convenience we assume that σ is irreducible. It will be seen that the final formula is independent of this assumption. We proceed as follows: Let $A(H)$ be the group algebra of H over the field of representation F. Then $A(H)$ is a subalgebra of $A(G)$ the group algebra of G. Let the minimal right ideal $\mathfrak{A}$ of $A(H)$ be the representation module for σ. Suppose that $a_1, a_2, \dots, a_r$ is a basis of $\mathfrak{A}$ over F. Form the rs elements

$$a_i t_j, \qquad i = 1, \dots, r; \quad j = 1, \dots, s \tag{23.1}$$

where $t_1 = 1$, $t_2, \ldots, t_s$ are right coset representatives in the decomposition

$$(23.2) \qquad G = Ht_1 + Ht_2 + \cdots + Ht_s .$$

Now the elements (23.1) are linearly independent over F, for if

$$\sum_{j=1}^{s} \left(\sum_{i=1}^{n} \lambda_{ij} a_i \right) t_j = 0, \qquad \lambda_{ij} \in F,$$

then each summand $(\sum_{i=1}^{r} \lambda_{ij} a_i) t_j = 0$, since the terms in the different summands belong to different cosets and so cannot cancel. But then

$$\left(\sum_{i=1}^{r} \lambda_{ij} a_i \right) t_j = 0 = \left(\sum_{i}^{r} \lambda_{ij} a_i \right) t_j t_j^{-1} = \sum_{i}^{r} \lambda_{ij} a_i .$$

Therefore $\lambda_{ij} = 0$, since the a_j are linearly independent. We assert that the set (23.1) is the basis of a right ideal $\mathfrak{A}^*$ in $A(G)$. It suffices to show that for any $g \in G$, $a_i t_j g$ is a linear combination of the elements of the set. Now from (23.2), $t_j g = ht_k$ for some unique $h \in H$ and some unique t_k so that

$$(23.3) \qquad a_i t_j g = a_i h t_k = \left(\sum_{j=1}^{r} f_{ij}^h a_j \right) t_k = \sum_{j=1}^{r} f_{ij}^h a_j t_k , \qquad f_{ij}^h \in F.$$

Thus $\mathfrak{A}^*$ is a right ideal of $A(G)$. It is then a representation module for a representation σ^* of G.

If we write the set (23.1) in s blocks: $a_1 t_1, \ldots, a_r t_1$; $a_1 t_2, \ldots, a_r t_2$; $\ldots$; $a_1 t_s, \ldots, a_r t_s$ we see from (23.3), with $i = 1, \ldots, r$, that the $rs \times rs$ matrix $\sigma(g)$ can be viewed as an $s \times s$ matrix with entries which are $r \times r$ matrices and that at the intersection of the jth row and kth column of $\sigma^*(g)$ there stands the $r \times r$ matrix $\sigma(h) = (f_{ij}^h)$. Other entries along the jth row of $\sigma^*(g)$ are then just the $r \times r$ zero matrix. Recalling the connection between g, h, j, k, viz., $t_j g t_k^{-1} = h$, we see that the matrix at the intersection of the jth row and lth column of $\sigma^*(g)$ is the $r \times r$ zero matrix, if $t_j g \notin H t_l$ but is $\sigma(t_j g t_l^{-1})$ if $t_j g t_l^{-1} \in H$. In summary we have

$$\sigma^*(g) = \begin{pmatrix} \sigma'(t_j g t_k^{-1}) \cdots \end{pmatrix} \cdots j\text{th row} \qquad (k\text{th column}) \tag{23.4}$$

where

$$\sigma'(t_j g t_k^{-1}) = \sigma(t_j g t_k^{-1}), \qquad \text{if} \qquad t_j g t_k^{-1} \in H$$
$$= 0, \text{ the } r \times r \text{ zero matrix, otherwise.}$$

If we adopt (23.4) as the definition of $\sigma^*(g)$ irrespective of whether σ is irreducible or not, it is a simple matter of matrix multiplication to show that $\sigma^*(g_1 g_2) = \sigma^*(g_1)\sigma^*(g_2)$. Thus $\sigma^*(g)$ is a representation of G if σ is any representation of H.

Let χ and χ^* be the characters of σ and σ^*, respectively. By (23.4)

$$\chi^*(g) = \sum_{j=1}^{s} \text{trace } \sigma'(t_j g t_j^{-1}) = \sum_{j=1}^{s} \chi'(t_j g t_j^{-1})$$

where $\chi'(x) = \chi(x)$ or 0 according as $x \in H$ or $x \notin H$. Since $h t_j g t_j^{-1} h^{-1} \in H \Leftrightarrow t_j g t_j^{-1} \in H$ and since

$$\chi(h t_j g t_j^{-1} h^{-1}) = \chi(t_j g t_j^{-1})$$

we have

$$\chi^*(g) = (1/H:1) \sum_{j=1}^{s} \sum_{h \in H} \chi'(h t_j g t_j^{-1} h^{-1}), \qquad H:1 = \text{order of } H.$$

Because every element $x \in G$ is realized uniquely as an $h t_j$, we get

(23.5)

$$\chi^*(g) = (1/H:1) \sum_{x \in G} \chi'(x g x^{-1}), \quad \text{and} \quad \chi'(x) = \chi(x), \quad x \in H,$$
$$= 0, \quad x \notin H.$$

This formula is independent of the choice of coset representatives t_j. Since a representation is determined by its character we see that σ^* as defined in (23.4) does not depend (up to equivalence) on the choice of the coset representatives of H in G. The result

(23.5) can be put in a different form. As x runs through the elements of G there will be repetitions: in fact, for each particular x_0, $x_0 g x_0^{-1}$ will occur $\mathfrak{N}(g) : 1$ times, that is, as often as the order of the normalizer $\mathfrak{N}(g)$ of g [if $b \in \mathfrak{N}(g)$: $(x_0 b) g (x_0 b)^{-1} = x_0(bgb^{-1})x_0^{-1} = x_0 g x_0^{-1}$, and conversely, if the first and last expressions are equal b permutes with g and thus $b \in \mathfrak{N}(g)$]. We now have

$$\chi^*(g) = (\mathfrak{N}(g) : 1/H : 1) \sum \chi(g')$$

where the sum is over elements g' similar to g but in H. If h_g denote the number of elements in the conjugate class of g then $h_g = G : 1/\mathfrak{N}(g) : 1$, so that if $m = G : 1/H : 1$ is the index of H in G, we can write

(23.6) $$\chi^*(g) = (m/h_g) \sum_{g' \in H \cap C_g} \chi(g'), \qquad C_g = \text{class of elements conjugate to } g$$

(23.7) **Theorem** (Frobenius reciprocity theorem). *Let ρ and σ be absolutely irreducible representations of a group G and a subgroup H, respectively. Then σ^*, the representation induced in G by σ, contains ρ with the same multiplicity with which σ occurs in ρ_* (i.e., ρ restricted to H).*

Proof. Let us recall (Remark, Section 20) that if χ is any character and χ^α an irreducible character of a group G, then

$$a = (1/G : 1) \sum_{g \in G} \chi(g)\chi^\alpha(g^{-1})$$

gives the multiplicity with which χ^α occurs in χ. Denote by χ^*, χ the characters of σ^* and σ, respectively, and let ψ be the character of ρ. Let a be the multiplicity of ρ in σ^* and let b be that of σ in ρ^*. Then

$$a = (1/G : 1) \sum \chi^*(g)\psi(g^{-1}),$$

and using (23.5)

$$a = (1/G : 1) \sum_{g \in G} \left((1/H : 1) \sum_{x \in G} \chi'(xgx^{-1})\right) \psi(g^{-1})$$

Since $\psi(xg^{-1}x^{-1}) = \psi(g^{-1})$,

$$a = (1/H : 1)(1/G : 1) \sum_{x \in G} \sum_{y \in G} \chi'(xgx^{-1})\psi(xg^{-1}x^{-1}).$$

For each fixed x the inner sum runs through all elements of G and since $\chi'(y) = 0$ if $y \notin H$ we have

$$a = (1/G : 1) \sum_{x \in G} \Big((1/H : 1) \sum_{h \in H} \chi(h)\psi(h^{-1})\Big).$$

Here the inner sum gives the multiplicity b of the irreducible character χ in ψ restricted to H. Thus

$$a = (1/G : 1) \sum_{x \in G} b = (1/G : 1)b(G : 1) = b$$

and the theorem is proved.

Remarks. There are a great number of interesting results on induced characters in the literature. (See, for example Curtis and Reiner [10], Chapter VI and references there.) We shall have a few more results later in this book but reasons of space prevent any extended treatment.

Since it is a simple matter to find the characters of a cyclic group it would be desirable if, for example, every character of a group G could be obtained as a linear combination with integral coefficients of characters induced by cyclic subgroups. This is not possible but the strongest result in this direction is the following theorem of Brauer.

Brauer's Theorem on Induced Characters. *Every character of G is a rational integral combination of characters induced from linear characters of elementary subgroups of G.*

A linear character is the character of a representation of degree one, and an elementary subgroup of G is one which is a direct product of a cyclic group and a p-group for some prime p.

If a group G has a pair of subgroups R and C with linear characters θ, ϕ, respectively, such that the induced characters θ^*, ϕ^* of G have exactly one irreducible character in common, each with multiplicity one, then this irreducible character can be calculated using the formula

$$\chi(g) = (n/[i(R \cap C : 1)]) \sum \theta(r)\phi(c)$$

where i is the number of elements in C_g, the conjugate class of g, n is the degree of the irreducible representation, and the summation is over all $r \in R$, $c \in C$ for which $rc \in C_g$ [8].

For the symmetric group S_n every irreducible character is accessable in this way from a pair of subgroups [see Section 28]. It is this result which underlies the method of the *Young tableau*: the permuted symbols 1, 2, ..., n when written in rows of nonincreasing length form a *shape*, or *tableau*. If we now take for R the subgroup of S_n permuting only the symbols within the rows and for C the subgroup permuting only those within the columns then this pair, with $\theta = 1$ on R and $\phi = \pm 1$ on the even or odd permutations of C, yield an irreducible character of S_n. Each shape yields such a pair and there are as many shapes as irreducible characters. For example, for S_3 there are the tableaux

1	2	3

1	2
3	

1
2
3

with the pairs $R = S_3$, $C = 1$; $R = \{1, (12)\}$, $C = \{1, (13)\}$, and $R = 1$, $C = S_3$, respectively.

A tableau is called a *standard tableau* if the numbers in each row and each column increase in magnitude. It can be shown that the degree of the irreducible character associated with a given shape is equal to the number of standard tableau of that shape (Rutherford [20], p. 68).

In the example given above all three shapes are standard tableaux but the second shape has also the standard tableau

1	3
2	

Hence it is associated with an irreducible character of degree 2.

These results make routine the construction of the character table of a symmetric group.

EXERCISES

1. Construct the character table of the symmetric group S_4.

2. Let G be a group and H a subgroup. Let χ be any character of G and ψ any character of H. Prove that (see Definition 20.2)

$$(\chi, \psi^G)_G = (\chi_H, \psi)_H$$

where $\chi_H = \chi \mid H$ and ψ^G is the character of G induced by ψ. [When ψ, χ are irreducible this is Frobenius theorem.]

3. Let $G \supset K \supset H$. Prove that $(\psi^K)^G = \psi^G$, if ψ is a character of H. [*Hint*: show that both sides contain an irreducible character χ equally often. Use Exercise 2.]

4. Let N be a normal subgroup of G and let ψ be an irreducible character of N. Let $T = \{g \in G : \psi(g^{-1}ng) = \psi(n), \ \forall\, n \in N\}$. (We say that T fixes ψ or that ψ is invariant under T.) Show that T is a group. Prove that $\psi^G \mid N$ contains ψ with multiplicity $T : N$. Show that if $T = N$, then ψ^G is irreducible.

5. Can the group ring of a nontrivial group be simple?

6. Let E be a finite extension field of degree d over F. Show that E is isomorphic to a matrix algebra A over F and that every automorphism of E which leaves F fixed corresponds to a transformation of A by a matrix T. [*Hint*: A is simple and hence has only one representation up to equivalence.]

7. Let G be a group and let M and N be F-$A(G)$ modules which afford representations μ and ν, respectively. Let π be any linear mapping (F-mapping) of M into N. Then $\mu(g)\pi\nu(g^{-1})$ is a linear mapping of M into N. Prove that $\alpha = \Sigma_{g \in G}\, \mu(g)\pi\nu(g^{-1})$ is an operator homomorphism (F-$A(G)$ mapping) of M into N.

8. Deduce from Exercise 7 that if $\mu(x) = (\mu_{ij}(x))$ and

$\nu(x) = (\nu_{jk}(x))$ are nonequivalent irreducible matrix representations then

$$\sum_{x \in G} \mu_{ij}(x)\nu_{rs}(x^{-1}) = 0,$$

$$\sum_{x \in G} \mu_{ij}(x)\mu_{rs}(x^{-1}) = n\delta_{is}\delta_{jr}/d,$$

where n denotes the order of the group G, d is the degree of the representation μ, and δ_{ab} are Kronecker deltas. [This result is also known as Schur's lemma. *Hint*: take $\pi = \pi_{ij}$ where for each basis element $m_k \in M$, $m_k\pi_{ij} = m_j\delta_{ki}$ and use (11.7) and (11.9).]

9. Deduce the character relation (19.12) from the results in Exercise 8.

10. Let $E = n/G : 1 \sum_{x \in G} x\, \chi(x^{-1})$, where χ is a character of degree n of the finite group G. Let $A(G)$ denote the group algebra of G over the complex field F. Show that E is in the center of $A(G)$. If χ is irreducible prove that $E^2 = E$ and that E is the identity of the minimal two-sided ideal containing the minimal right ideal of $A(G)$, which provides the representation with character χ.

CHAPTER IV

Applications of the Theory of Characters

24. Algebraic Numbers

Any root $x = \alpha$ of an equation,

$$x^n + a_1 x^{n-1} + \cdots + a_{n-1}x + a_n = 0, \tag{24.1}$$

in which the a_i are rational numbers is called an *algebraic number*.

If the coefficients a_i are rational integers the algebraic number is called an *algebraic integer*.

We shall need a few facts concerning algebraic numbers and algebraic integers. These will be proved in the lemmas of this section.

(24.2) **Lemma.** *An algebraic integer which is a rational number is a rational integer.*

Proof. Let $\alpha = r/s$, r, s rational integers with no common divisor. Suppose that α satisfies (24.1) in which a_i are rational integers. Then

$$r^n = -(a_1 r^{n-1} s + \cdots + a_n s^n).$$

Since s is a factor on the right side of this equation any prime p dividing s divides r^n and hence r.

But r and s are relatively prime and hence $s = 1$, and α is a rational integer as required.

(24.3) **Lemma.** *Let $\alpha, \beta, \ldots, \lambda$ be algebraic numbers (integers). A polynomial $f(\alpha, \beta, \ldots, \lambda)$ with coefficients which are rational numbers (integers) is itself an algebraic number (integer).*

Proof. Let α be a root of $A(x) = 0$ of degree a, β a root of $B(x) = 0$ of degree b, ..., and λ a root of $L(x) = 0$ of degree l where the coefficients of all equations are rational numbers (integers) and the leading coefficient is unity.

If $N = ab \cdots l$ form the N numbers

$$\alpha^{a'}\beta^{b'} \cdots \lambda^{l'},$$

$$0 \leqslant a' \leqslant a - 1, \qquad 0 \leqslant b' \leqslant b - 1, \ldots, 0 \leqslant l' \leqslant l - 1,$$

which, arranged in some fixed order, we denote by $\omega_1, \omega_2, \ldots, \omega_N$.

Using $A(\alpha) = 0$, α^a and higher powers of α can be expressed as polynomials in α of degree $<a$ with coefficients which are rational numbers (integers). Similarly for the other polynomials. Hence $\omega_i f$ can be expressed in the form

$$\omega_i f = c_{i1}\omega_1 + c_{i2}\omega_2 + \cdots + c_{iN}\omega_N, \qquad i = 1, 2, \ldots, N, \tag{24.4}$$

in which each c_{ij} is a polynomial in the coefficients of $f, A, \ldots, L$ and as such are therefore rational numbers (integers).

But (24.4) is the system of linear homogeneous equations for the ω_i:

$$\begin{aligned}
(c_{11} - f)\omega_1 + \quad c_{12}\omega_2 + \cdots + \quad c_{1N}\omega_N &= 0\\
c_{21}\omega_1 + (c_{22} - f)\omega_2 + \cdots + \quad c_{2N}\omega_N &= 0\\
\cdots \quad &\\
c_{N1}\omega_1 + \quad c_{N2}\omega_2 + \cdots + (c_{NN} - f)\omega_N &= 0
\end{aligned}$$

Since the ω_i are not all zero, the determinant of the coefficients must vanish. Multiplying it by $(-1)^N$ we get an equation,

$$f^N - (c_{11} + c_{22} + \cdots + c_{NN})f^{N-1} + \cdots = 0,$$

whose coefficients are rational numbers (integers). Thus f is an algebraic number (integer).

(24.5) **Corollary.** *The set of algebraic integers form a ring.*

Proof. $\alpha \pm \beta$ and $\alpha\beta$ are algebraic integers if α and β are algebraic integers.

(24.6) **Corollary.** *The set of algebraic numbers is a field.*

Proof. $\alpha \pm \beta$ and $\alpha\beta$ are algebraic numbers if α and β are. Besides, if $\alpha \neq 0$ satisfies the equation $f(x) \equiv x^n + a_1x^{n-1} + \cdots + a_n = 0$ with a_i rational numbers, then $1/\alpha$ satisfies the equation $a_nx^n + \cdots + a_1x + 1 = 0$ with rational coefficients.

It is clear that any rational integer n is an algebraic integer, since it satisfies the equation $x - n = 0$. If α and γ are two algebraic integers, we say that α divides γ, written $\alpha \mid \gamma$, or that α is a divisor of γ if there is an algebraic integer β such that $\gamma = \alpha\beta$.

(24.7) **Lemma.** *Let a and b be rational integers and let γ be an algebraic integer. If $b \mid a\gamma$ and a and b are relatively prime, then $b \mid \gamma$.*

Proof. Since a and b are relatively prime there are rational integers c and d such that $ac + bd = 1$. Hence $a\gamma c + b\gamma d = \gamma$. But $a\gamma = \beta b$ for some algebraic integer β. Thus $b(\beta c + \gamma d) = \gamma$, or $b \mid \gamma$.

25. Some Results from the Theory of Characters

In (19.3) we had

$$f_t^i = (h_i/n_t)\chi_i^t$$

where χ_i^t is the value of the irreducible character χ^t on any element of the conjugate class C_i of group elements, h_i is the number of elements in C_i, and n_t is the degree of the representation ρ_t, of which χ^t is the character.

Moreover we saw (19.5) that

$$f_t^i f_t^j = \sum_{k=1}^{s} l_k^{ij} f_t^k,$$

l_k^{ij} are nonnegative integers and $i = 1, 2, \ldots, s$, where s is the number of conjugate classes. For fixed j and $i = 1, 2, \ldots, s$ this gives the system of linear homogeneous equations

$$\sum_{k=1}^{s} (l_k^{ij} - \delta_{ki} f_t^j) f_t^k = 0$$

for the f_t^k. Since these are not all zero [$\rho_t \neq 0$, (19.1)] the determinant of coefficients must vanish. Explicitly:

$$\begin{vmatrix} l_1^{1j} - f_t^j & l_2^{1j} & \cdots & l_s^{1j} \\ l_1^{2j} & l_2^{2j} - f_t^j & \cdots & l_s^{2j} \\ \cdot & \cdot & \cdots & \cdot \\ l_1^{sj} & l_2^{sj} & \cdots & l_s^{sj} - f_t^j \end{vmatrix} = 0.$$

Expanding and then multiplying through by $(-1)^s$ we get

$$(f_t^j)^s - (l_1^{1j} + l_2^{2j} + \cdots + l_s^{sj})(f_t^j)^{s-1} + \cdots 1 = 0.$$

Thus f_t^j satisfies an equation with rational integral coefficients and leading coefficient unity. Hence:

(25.1) **Lemma.** *The numbers $(h_i/n_t)\chi_i^t$ are algebraic integers.*

Let σ be a representation of degree f of the finite group G. Any element $g \in G$ has some finite order m. Thus $g^m = 1$ and g generates a cyclic subgroup H of order m. Now σ restricted to H is completely reducible and all its irreducible constituents are of degree one (Corollary 18.3). Hence there is a matrix T such that

$$T^{-1}\sigma(g)T = \begin{pmatrix} \lambda_1 & & & \\ & \lambda_2 & & \\ & & \ddots & \\ & & & \lambda_f \end{pmatrix}.$$

Raising to the power m we get the identity matrix on the left and thus $\lambda_i^m = 1$ so that the λ_i are algebraic integers. Now $\chi^\sigma(g) = \sum_{i=1}^{f} \lambda_i$ is also an algebraic integer, so that we can state:

(25.2) **Lemma.** *The value of any character on a group element of a finite group is an algebraic integer.*

(25.3) **Theorem.** *The degree n_t of an absolutely irreducible representation ρ_t of a finite group is a divisor of the order n of the group.*

Proof. Using the orthogonal relation (19.11) with $i = j = t$ we get

$$\sum_{a=1}^{s} [(h_a/n_t)\chi^t(g_a)][\chi^t(g_a^{-1})] = n/n_t\,.$$

By Lemmas 25.1 and 25.2 each term in square brackets is an algebraic integer and hence the left-hand side is an algebraic integer. As n and n_t are rational integers the right side is a rational number and so, by Lemma 24.2, is a rational integer.

As we have seen, $(h_i/n_t)\chi_i^t$ is an algebraic integer. If the rational integers h_i, n_t are relatively prime Lemma 24.7 shows that χ_i^t/n_t is an algebraic integer. Now, as seen in the proof of Lemma 25.2,

$$\chi_i^t = \omega^{m_1} + \omega^{m_2} + \cdots + \omega^{m_{n_t}}$$

where ω is an mth root of unity for some m. Thus

$$|\,\chi_i^t\,| \leqslant \sum_{j=1}^{n_t} |\,\omega^{m_j}\,| = n_t\,,$$

and, for example, thinking of the ω^{m_j} as unit vectors in the complex plane, it is seen that equality holds if and only if

$$\omega^{m_1} = \omega^{m_2} = \cdots = \omega^{m_{n_t}} = \omega.$$

Hence

(25.4) either $|\,\chi_i^t/n_t\,| < 1$ or $\chi_i^t = n_t\omega.$

Write

$$\alpha = \chi_i^t/n_t = \sum_{j=1}^{n_t} \omega^{m_j}\Big/n_t\,,$$

and α', α'', .., $\alpha^{(m-1)}$ for α, when the ω in the right-hand side is replaced by its conjugates ω^2, ω^3, ..., ω^m, respectively. Then the polynomial

$$f(x) = (x - \alpha)(x - \alpha') \cdots (x - \alpha^{(m-1)}) \tag{25.5}$$

has for its coefficients symmetric functions (with rational coefficients) of ω, ω^2, ..., $\omega^m = 1$. Thus the coefficients of $f(x)$ are rational numbers.

Now let

$$g(x) = x^r + c_1 x^{r-1} + \cdots + c_r = 0 \tag{25.6}$$

be the irreducible equation with integral coefficients satisfied by the algebraic integer α. The roots of this equation are called conjugates of α. Since α is a root of (25.5) and (25.6), $g(x)$ is a divisor of $f(x)$ and so the conjugates of α are among the roots α', ..., $\alpha^{(m-1)}$ of $f(x)$. Hence for the conjugates of α, denoted by α_2, ..., α_r, there holds the same inequality:

$$|\alpha_j| \leqslant 1.$$

But $\alpha\alpha_2 \cdots \alpha_r = \pm c_r$, a rational integer. If $|\alpha| < 1$, we have

$$|c_r| = |\alpha\alpha_2 \cdots \alpha_r| = |\alpha||\alpha_2| \cdots |\alpha_r| < 1$$

and so

$$c_r = 0.$$

Since (25.6) is the irreducible equation for α this implies that $\alpha = 0$.

Returning to (25.4), since $\alpha = \chi_i^t/n_t$, we state:

(25.7) **Lemma.** *If the degree n_t of an irreducible representation ρ_t of a finite group G is prime to h_i, the number of elements in the conjugate class C_i, then either $\chi_i^t = 0$ or $\chi_i^t = n_t\omega$, where ω is a root of unity. In the latter case $\rho_t(C_i)$ lies in the center of $\rho_t(G)$.*

Proof. Only the final statement remains to be proved. In this case $|\chi_i^t| = n_t$ and the result follows from the more general:

(25.8) **Lemma.** *Let μ be any representation of degree m of a finite group G. If for some $g \in G$*

$$|\chi^{\mu}(g)| = m,$$

then $\mu(g)$ is a scalar matrix:

$$\mu(g) = \begin{pmatrix} \omega & & 0 \\ & \ddots & \\ 0 & & \omega \end{pmatrix}, \qquad \omega \text{ a root of unity.}$$

If $\chi^{\mu}(g) = m$, then $\mu(g) = I$, the identity matrix, and hence g belongs to the kernel of μ.

Proof. Let $g^k = 1$. Form the subgroup $H = \{1, g, ..., g^{k-1}\}$. Since H is an abelian group its irreducible representations $\sigma_1, \sigma_2, ..., \sigma_k$ are of degree 1 (Corollary 18.3). Now μ restricted to H is a representation and is completely reducible (Corollary 16.5). Hence there is a matrix T such that

$$T^{-1}\mu(g)T = \begin{pmatrix} \sigma_{j_1}(g) & & 0 \\ & \ddots & \\ 0 & & \sigma_{j_m}(g) \end{pmatrix}$$

and the $\sigma_j(g)$ are complex numbers. Since $g^k = 1$, $\sigma_{j_1}(g), ..., \sigma_{j_m}(g)$ are roots of unity. Then

$$\chi^{\mu}(g) = \sigma_{j_1}(g) + \cdots + \sigma_{j_m}(g)$$

and so

$$(*) \qquad |\chi^{\mu}(g)| \leqslant |\sigma_{j_1}(g)| + \cdots + |\sigma_{j_m}(g)| = m.$$

Now equality can hold in $(*)$ if and only if

$$\sigma_{j_1}(g) = \sigma_{j_2}(g) = \cdots = \sigma_{j_m}(g) = \omega,$$

say. Hence

$$T^{-1}\mu(g)T = \omega I, \qquad I \text{ the identity matrix.}$$

But then

$$\mu(g) = \omega I.$$

Hence

$$\chi^{\mu}(g) = m\omega.$$

Then if $\chi^{\mu}(g) = m$, we have $\omega = 1$ and

$$\mu(g) = I.$$

The proof is now complete.

26. Normal Subgroups and the Character Table

A. The Existence of Normal Subgroups

Lemma 25.8 shows that a simple inspection of the character table of a finite group can reveal the presence of a normal subgroup. Thus if in the accompanying table the first number

	$C_1 = 1$	C_2	$\cdots$	C_i	$\cdots$	C_s
χ^1	1	1	$\cdots$	1	$\cdots$	1
χ^2	n_2	χ_2^2	$\cdots$	χ_i^2	$\cdots$	χ_s^2
$\vdots$						
χ^α	n_α	χ_2^α	$\cdots$	χ_i^α	$\cdots$	χ_s^α
$\vdots$						
χ^s	n_s	χ_2^s	$\cdots$	χ_i^s	$\cdots$	χ_s^s

in any row, other than the first, is the same as some other number in that row, say $\chi^\alpha(1) = n_\alpha = \chi^\alpha(g_i)$, $g_i \in C_i$, then g_i $(\neq 1)$ lies in the kernel of the corresponding representation. Hence there is a nontrivial normal subgroup. Moreover, this normal subgroup consists precisely of all the elements g satisfying $\chi^\alpha(g) = \chi^\alpha(1) = n_\alpha$.

That the presence of a normal subgroup is invariably detected in this way is shown in the proof of the following lemma. Recall that a group is simple if it has no normal subgroups.

(26.1) **Lemma.** *A group G is simple if and only if $\forall g \in G$ and for every irreducible character χ^α, not the 1-character, $\chi^\alpha(g) \neq \chi^\alpha(1)$ if $g \neq 1$.*

Proof. (1) Lemma 25.8 shows that if G is simple then the condition holds.

(2) Now suppose that $N \lhd G$. Let $\hat{\sigma}$ be an irreducible representation of G/N. Define $\sigma(g) = \hat{\sigma}(\bar{g})$, where $\bar{g}$ is the class of g modulo N. Since

$$\sigma(gg') = \hat{\sigma}(\overline{gg'}) = \hat{\sigma}(\overline{gg'}) = \hat{\sigma}(\bar{g})\hat{\sigma}(\bar{g}') = \sigma(g)\sigma(g'),$$

therefore σ is a representation of G, and, thinking in terms of matrices, it is easy to see that σ is irreducible, as G and G/N are represented by the same matrices. Now if g belongs to the kernel of σ (which contains N) we have $\chi^\sigma(g) = \chi^\sigma(1)$ and the condition does not hold. This proves the lemma.

B. The Determination of All Normal Subgroups

We have seen that inspection of the character table of a group G gives at once a normal subgroup which is the kernel of an irreducible representation. Now the proof of Lemma 26.1 shows that if G has any normal subgroup N then it has some normal subgroup N' which is the kernel of an irreducible representation. However, N itself may not be immediately accessible from the table. To find all normal subgroups we can proceed as follows: if $n = G : 1$, for each $m \mid n$ let us form a character χ,

$$\chi = a_1\chi^1 + \cdots + a_s\chi^s \tag{26.2}$$

where the a_i are integers $\geqslant 0$ and such that

$$m = a_1\chi^1(1) + \cdots + a_s\chi^s(1). \tag{26.3}$$

Then $N_m^\chi = \{g : \chi(g) = \chi(1)\}$ is a (possibly trivial) normal subgroup. Because of (26.3) there is only a finite number of χ to be examined for each m.

We must show that any normal subgroup N is obtained in this way. Let $\hat{\rho}$ be the regular representation (or any faithful representation) of G/N. Then $\hat{\rho}$ is faithful and its degree is $G : N$. As in Section A, if $\rho(g) = \hat{\rho}(\bar{g})$, then ρ is a representation of G. Now the kernel of ρ is precisely N and

$$\chi^\rho(g) = \chi^\rho(1) \qquad \text{if and only if} \quad g \in N.$$

But as a character of G, χ^ρ is realized by (26.2). Hence N will be obtained in the process described.

27. Some Classical Theorems

The following theorems of Burnside and Frobenius have been for a long time showpieces of the application of the theory of characters to group theory.

(27.1) **Theorem** (Burnside). (1) *If the number h_i of elements in some conjugate class C_i of a group of finite order is a power of a prime, the group cannot be simple.*

(2) *A group of order $p^a q^b$, p and q primes, is solvable.*

Proof. (1) Let $h_i = p^m$, $g \in C_i$. If n is the order of the group G and $n_1 = 1, n_2, \ldots, n_s$ the degrees of the irreducible representations of G we have

$$n = 1^2 + n_2^2 + \cdots + n_s^2 .$$

Since $p \mid n$, p cannot divide all n_i, hence there is an n_t relatively prime to p^m. Then by Lemma 25.7

$$\chi_i^t = \chi^t(g) = 0 \qquad \text{or} \qquad \rho_t(C_i) \text{ is in the center of } \rho_t(G).$$

In the latter case if N' denote the center of $\rho_t(G)$ and if $N = \{x : x \in G, \quad \rho_t(x) \in N'\}$, then $N \supset C_i$ and N is a normal subgroup of G. Hence G is not simple. Thus, if we assume G simple, for each $(n_\alpha , p) = 1$ we have $\chi^\alpha(g) = 0$. From the first character relation (19.9)

$$(*) \qquad \sum_{\alpha=1}^{s} \chi^\alpha(1)\chi^\alpha(g) = 0.$$

But since $\chi^\alpha(1) = n_\alpha$ and $\chi^1(g) = 1$, there is a summand 1 in $(*)$ and any summand prime to p is zero. Hence

$$1 + \cdots \equiv 0 \qquad (\text{mod}\, p)$$

Since this is impossible G is not simple.

(2) Let H be a Sylow p-subgroup of G. Then $H : 1 = p^a$. Now H, as a p-group, has a nontrivial center $\mathfrak{Z}(H)$. Let $1 \neq g \in \mathfrak{Z}(H)$. Recall that the number of elements h_0 in the class of g is equal to the index of the normalizer $\mathfrak{N}(g)$ in G. But

$$\mathfrak{N}(g) \supset H, \qquad \text{since} \qquad g \in \mathfrak{Z}(H).$$

$$\therefore \quad \mathfrak{N}(g) : 1 \geqslant p^a.$$

$$\therefore \quad h_g = G : 1/\mathfrak{N}(g) : 1 = p^a q^b/\mathfrak{N}(g) : 1 = q^{b'}, \qquad b' \leqslant b.$$

Hence the number of elements in the class of g is a power of a prime. Then G is not simple by the first part and has a normal subgroup N. Since N and G/N are groups of lower order but of order $p^\alpha q^\beta$, an induction on the order of G yields the result that G is solvable.

Recall that a permutation group is *transitive* if for every pair of symbols there is a permutation which takes any one of them into the other.

(27.2) **Theorem** (Frobenius). *If G is a transitive permutation group on n symbols whose permutations other than the identity leave at most one of the symbols fixed, then those permutations of G which displace all the symbols, together with the identity, form a normal subgroup of order n.*

Proof. Let G permute 1, 2, ..., n and let t_i, $i = 1, 2, ..., n$, be elements of G such that $1 \cdot t_i = i$. Then if $H = H_1$ is the subgroup leaving 1 fixed, $G = H + Ht_2 + \cdots + Ht_n$ is a coset decomposition of G. Now if H_i keeps i fixed, we have $H_i = t_i^{-1}Ht_i$, $i = 1, 2, ..., n$, and so all H_i have the same order $H : 1$. By hypothesis $H_i \cap H_j = 1$, $i \neq j$. By easy calculation: $\bigcup_{i=1}^{n} H_i : 1 = (H : 1 - 1)n + 1$, $G : 1 = (H : 1)n$, so that the number of elements outside of $\bigcup H_i$, which displace all symbols, is $n - 1$. We wish to prove that these elements together with the identity form a normal subgroup of G.

The idea of the proof is to construct representations of G which have these elements as the intersection of their kernels. Consider the vector space of dimension n with a basis $X_1, X_2, ..., X_n$. Let $\sigma(g)$, $g \in G$, be the linear transformation given by

$$X_i\sigma(g) = X_{i \cdot g} .$$

Then, if χ^σ is the character of the representation σ,

$$\chi^\sigma(1) = n,$$

$$\chi^\sigma(x) = 0, \qquad x \notin \bigcup H_i, \qquad \text{since } i \cdot x \neq i,$$

$$\chi^\sigma(y) = 1, \qquad 1 \neq y \in \bigcup H_i, \qquad \text{since } \forall y, \exists \text{ exactly one } j: j \cdot y = j.$$

Now if χ^1 is the one-character:

$$(1/G : 1) \sum_{g \in G} \chi^\sigma(g)\chi^1(g^{-1}) = (1/G : 1)\Big[\chi^\sigma(1) \cdot 1 + \sum_{1 \neq y \in \bigcup H_i} \chi^\sigma(y) \cdot 1\Big]$$
$$= (1/G : 1)[n + (H : 1 - 1)n] = 1,$$

Hence χ^1 is contained once in χ^σ, and so $\chi = \chi^\sigma - \chi^1$ is a character. Then

$$\begin{aligned} &\chi(1) = n - 1, \\ (27.3) \qquad &\chi(x) = -1, \qquad x \notin \bigcup H_i, \\ &\chi(y) = 0, \qquad 1 \neq y \in \bigcup H_i . \end{aligned}$$

Now let ψ^α be an arbitrary irreducible character of H, and let $\psi^{*\alpha}$ be the corresponding induced character of G.

Using formula (23.5) we calculate

$$
\begin{aligned}
&\psi^{*\alpha}(1) = n\psi^{\alpha}(1), && \\
(27.4)\qquad &\psi^{*\alpha}(x) = 0, && x \notin \bigcup H_i, \\
&\psi^{*\alpha}(y) = \psi^{\alpha}(y'), && y' \in H, \\
& && y' \text{ conjugate to } y \in \bigcup H_i .
\end{aligned}
$$

Form the generalized character ϕ^{α}:

$$\phi^{\alpha} = \psi^{*\alpha} - \psi^{\alpha}(1)\chi$$

and find

$$
\begin{aligned}
&\phi^{\alpha}(1) = \psi^{\alpha}(1), && \\
(27.5)\qquad &\phi^{\alpha}(x) = \psi^{\alpha}(1) = \phi^{\alpha}(1), && x \notin \bigcup H_i, \\
&\phi^{\alpha}(y) = \psi^{\alpha}(y'), && 1 \neq y \in \bigcup H_i, \\
& && y' \in H, \\
& && y' \text{ conjugate to } y.
\end{aligned}
$$

We will now show that ϕ^{α} is an irreducible character of G. Already $\phi^{\alpha}(1) > 0$, and

$$
\begin{aligned}
(\phi^{\alpha}, \phi^{\alpha}) &= 1/G:1 \sum_{g \in G} \phi^{\alpha}(g)\overline{\phi^{\alpha}(g)} \\
&= 1/G:1 \left[(\phi^{\alpha}(1))^2 + \sum_{x \notin \bigcup H_i} \phi^{\alpha}(x)\overline{\phi^{\alpha}(x)} + \sum_{1 \neq y \in \bigcup H_i} \phi^{\alpha}(y)\overline{\phi^{\alpha}(y)}\right] \\
&= 1/G:1\left[(\phi^{\alpha}(1))^2 + (\phi^{\alpha}(1))^2(n-1) + n \sum_{1 \neq y' \in H} \psi^{\alpha}(y')\overline{\psi^{\alpha}(y')}\right].
\end{aligned}
$$

Now because ψ^{α} is an irreducible character on H we have $(1/H:1)\ \Sigma_{y' \in H}\, \psi^{\alpha}(y')\overline{\psi^{\alpha}(y')} = 1$ so that the last summation in the bracket yields

$$H:1 - \psi^{\alpha}(1)\overline{\psi^{\alpha}(1)} = H:1 - (\phi^{\alpha}(1))^2.$$

$$
\begin{aligned}
\therefore\quad (\phi^{\alpha}, \phi^{\alpha}) &= (1/G:1)[(\phi^{\alpha}(1))^2 + (\phi^{\alpha}(1))^2(n-1) \\
&\qquad + n(H:1 - (\phi^{\alpha}(1))^2)] \\
&= (1/G:1)n(H:1) = 1.
\end{aligned}
$$

Thus ϕ^{α} is an irreducible character of G.

Since $\phi^\alpha(x) = \phi^\alpha(1)$, $x \notin \bigcup H_i$ we have by Lemma 25.8 that $G - \bigcup H_i$ belongs to the kernel of τ_α, the representation whose character is ϕ^α. Now this is true for each τ_α, corresponding to each irreducible character ψ^α on H. On the other hand, if $1 \neq y \in \bigcup H_i$ we cannot have $\phi^\alpha(y) = \phi^\alpha(1)$ for all τ_α, for then by (27.5) $\exists y' \in H$, $y' \neq 1$, such that $\psi^\alpha(y') = \psi^\alpha(1)$, for every α. But then we should have the element $1 - y'$ of the group algebra $A(H)$ of H mapped onto zero by every irreducible representation α of H. Thus $1 - y' \neq 0$ belongs to the radical of $A(H)$. As $A(H)$ is semisimple this is impossible. Hence $(G - \bigcup H_i) \bigcup 1$ is equal to the intersection of the kernels of the τ_α and as such is a normal subgroup of G. This proves the theorem.

EXERCISES

1. Show that a p-group has a linear character whose kernel is a subgroup N of index p. Prove that $\chi \mid N$ (χ restricted to N) is reducible.

2. Use the relation (16.7) and Theorem 25.3 to show that a group of order p^2 is abelian; also that a group of order p^n has a normal subgroup of order $\leqslant p^{n-1}$.

3. If χ is an irreducible character of a group G, not of degree one, show that $\chi(g) = 0$ for some element $g \in G$. (Burnside [7], p. 319, Ex. 5.)

CHAPTER V

The Construction of Irreducible Representations

28. Primitive Idempotents

Let A be the group algebra of a finite group G over the complex field F. We have seen from the structure theorems (Section 10) that there is, up to arrangement of summands, exactly one decomposition

$$A = A_1 \oplus \cdots \oplus A_s$$

in which the simple rings A_α are mutually annihilating simple two-sided ideals of A. Moreover, each A_α, called a block, is the direct sum of n_α isomorphic right ideals of A and is besides isomorphic to a complete matrix algebra of degree n_α over F. Thus there are n_α^2 elements $e_{jk}^\alpha \in A_\alpha$, $j, k = 1, \ldots, n_\alpha$, satisfying

$$e_{ij}^\alpha e_{kl}^\alpha = \delta_{jk} e_{il}^\alpha, \qquad \delta_{jk} \text{ the Kronecker delta,} \tag{28.1}$$

and such that any $a \in A_\alpha$ can be written uniquely as

$$a = \sum_{i,j=1}^{n_\alpha} f_{ij} e_{ij}^\alpha, \qquad f_{ij} \in F. \tag{28.2}$$

Furthermore, since the A_α are mutually annihilating

$$e_{ij}^\alpha e_{kl}^\beta = 0, \qquad \alpha \neq \beta, \tag{28.3}$$

it is easy to see from the relations (28.1) and (28.3) that

$$1 = \sum_{\alpha=1}^{s} \Big(\sum_{i=1}^{n_\alpha} e_{ii}^{\alpha} \Big), \qquad 1 \text{ the identity of } A,$$

and that the inner sum E_α is the *unique* identity element of A_α. Now $R_{\alpha j} = e_{jj}^{\alpha} A$, $j = 1, 2, \ldots, n_\alpha$, are right ideals of A and we have

$$A_\alpha = E_\alpha A_\alpha = E_\alpha \Big(\sum_{\beta=1}^{s} A_\beta \Big) = E_\alpha A = \Big(\sum_{j=1}^{n_\alpha} e_{jj}^{\alpha} \Big) A$$

$$= R_{\alpha 1} \oplus \cdots \oplus R_{\alpha n_\alpha}.$$

By the uniqueness of the Remak decomposition the $R_{\alpha j}$, $j = 1, \ldots, n_\alpha$, must be isomorphic minimal right ideals. Any one of them, say $R_{\alpha 1}$, is a representation module for an irreducible representation σ_α. It can be seen from the relations (28.1) and (28.3) that $e_{11}^{\alpha}, \ldots, e_{1n}^{\alpha}$ is an F-basis of $R_{\alpha 1}$.

An element e of an algebra is an *idempotent* if $e^2 = e$. It is a *primitive idempotent* if it cannot be written as $e = e_1 + e_2$ where $e_1^2 = e_1 \neq 0$, $e_2^2 = e_2 \neq 0$, and $e_1 e_2 = e_2 e_1 = 0$. If e is not primitive then the right ideal $eA = e_1 A \oplus e_2 A$, so that eA is not a minimal right ideal. Conversely, suppose that eA is not minimal. Then the representation module eA is reducible and, since A is semisimple, it is decomposable (Theorem 9.3) so that $eA = eA_1 \oplus eA_2$. Hence:

$$e = ea + ea', \qquad ea \in eA_1, \qquad ea' \in eA_2.$$

Now

$$e = e^2 = eae + ea'e = ea + ea'$$

which implies that

$$eae = ea, \qquad ea'e = ea'.$$

Also

$$ea = ea^2 + ea'a, \qquad \text{so that} \qquad ea = ea^2, \qquad ea'a = 0.$$

Thus

$$ea = eaa = (eae)a = (ea)^2$$

and

$$ea'a = (ea'e)a = (ea')(ea) = 0.$$

Similarly

$$(ea')^2 = ea' \qquad \text{and} \qquad (ea)(ea') = 0.$$

This shows that e is not primitive. We state this result as:

(28.4) **Lemma.** *An idempotent e of a semisimple algebra A is primitive if and only if the right ideal eA is minimal.*

This result shows that the elements e_{ii}^{α}, $i = 1, 2, ..., n_{\alpha}$, are primitive idempotents of the αth block and any one of them, say e_{11}^{α}, generates a minimal right ideal $e_{11}^{\alpha}A = R_{\alpha 1}$ which is a representation module for the irreducible representation σ_{α}. The problem of constructing irreducible representations can thus be reduced to that of finding a primitive idempotent for each block.

It has been pointed out (Remarks, Section 23) that for the symmetric group S_n each Young tableau is associated with an irreducible representation. Actually each shape gives a primitive idempotent

$$e = (n_{\alpha}/n!) \sum \phi(c)rc \tag{28.5}$$

where n_{α} is the degree of the corresponding irreducible representation, the summation is over all elements $r \in R$, $c \in C$, and $\phi(c) = \pm 1$ according to whether c is an even or odd permutation. This follows from a combinatorial lemma of von Neumann [van der Waerden [22], Vol. 2, p. 191], the precise generalization of which can be stated for an arbitrary group as:

(28.6) **Lemma.** *Let two subgroups R and C of a group G have representations θ and ϕ, respectively, both of the first degree. Suppose that the following condition holds $\forall s \in G$: $s \notin CR$ if and only if*

$\exists r \in R$, $c \in C$, such that $srs^{-1} = c$ and $\theta(r) \neq \phi(c)$. Then $e = PN$ is a multiple of a primitive idempotent, where

$$P = \sum_{r \in R} r\theta(r), \qquad N = \sum_{c \in C} c\phi(c).$$

Proof. First note that

$$Pr'\theta(r') = \sum_{r \in R} r\theta(r)r'\theta(r') = \sum_{r \in R} rr'\theta(rr') = P$$

where r' is any element of R. Similarly $Nc'\phi(c') = N$. Consider $PNsPN$. If $s \in CR$, $s = cr$ say, then

$$\begin{aligned} PNsPN = PNcrPN &= \theta^{-1}(r)\phi^{-1}(c)P(Nc\phi(c))(r\theta(r)P)N \\ &= \theta^{-1}(r)\phi^{-1}(c)(PN)^2. \end{aligned}$$

On the other hand, if $s \notin CR$, then by the condition of the lemma there is an r, c such that $srs^{-1} = c$ and $\theta(r) \neq \phi(c)$. Then

$$PNsPN = \theta(r)PNsrs^{-1}sPN = \theta(r)PNcsPN = \theta(r)\phi^{-1}(c)PNsPN.$$

Hence

$$PNsPN(1 - \theta(r)\phi^{-1}c) = 0.$$

Since $\theta(r) \neq \phi(c)$ we have $PNsPN = 0$. Writing $e = PN$ we get

$$eAe = Fe^2 \tag{28.7}$$

where A is the group algebra of G over F the complex field. Note that $e \neq 0$, for the coefficient of identity I in PN is $f = \Sigma\, \theta(r)\phi(c)$ where the summation is over all r, c for which $rc = I$, that is, over all $r \in R \cap C$. But $IrI^{-1} = c^{-1}$ and since $I \in CR$ the condition of the lemma gives $\theta(r) = \phi^{-1}(c)$. Hence $f = R \cap C : 1 \neq 0$. Also $e^2 \neq 0$ otherwise by (28.7) $eAeA = 0$ and eA is a nilpotent right ideal which is impossible because A is semisimple. By considering the expression $Ps^{-1}N$ and reasoning exactly as in the case $PNsPN$ we find that $PAN = FPN = Fe$. Since $PAN \supset PNAPN = eAe$, we get

$$Fe \supset eAe. \tag{28.8}$$

Equations (28.7) and (28.8) yield

$$(28.9) \qquad e^2 = fe \qquad \text{for some} \qquad f \in F$$

so that e is a multiple of an idempotent. Now if e is not a primitive idempotent then $e = e_1 + e_2$, where e_1, e_2 are nonzero idempotents and $e_1 e_2 = e_2 e_1 = 0$. But then $0 \neq e_1 = e e_1 e = f'e$, for some $f' \in F$, by (28.7) and (28.9) and so $0 = e_1 e_2 = f' e e_2 = f' e_2 \Rightarrow e_2 = 0$, a contradiction. Thus e is a primitive idempotent. This completes the proof.

Now let G be a group satisfying the conditions of Lemma 28.6. Then $e = f'PN = f' \sum_{r \in R, c \in C} \theta(r)\phi(c)rc$ is a primitive idempotent where f' is some element of the field F. Let us find an expression for f'. We know that the minimal right ideal eA must be isomorphic to some right ideal $e_{11}^{\alpha}A$. This implies that there is a unit $u \in A$ such that $u^{-1}eu = e_{11}$ [see Appendix]. Then if χ^{ρ} denotes the character of the regular representation, $\chi^{\rho}(e) = \chi^{\rho}(u^{-1}eu) = \chi^{\rho}(e_{11}^{\alpha})$. Now as a basis for the regular representation of A we can take the elements $\{e_{ij}^{\alpha} : i, j = 1, \ldots, n_{\alpha}; \quad \alpha = 1, \ldots, s\}$. Since $e_{ij}^{\alpha} e_{kl}^{\beta} = \delta_{\alpha\beta}\delta_{jk} e_{il}^{\alpha}$ we see that χ^{ρ}, the trace of the regular representation, is given by $\chi^{\rho}(e_{kl}^{\alpha}) = n_{\alpha}$, 0 according as $k = l$ or not. Thus $\chi^{\rho}(e) = \chi^{\rho}(e_{11}^{\alpha}) = n_{\alpha}$. But $\chi^{\rho}(e) = \chi^{\rho}(f'PN) = f'\chi^{\rho}(PN) = f'f(G : 1)$ where f is the coefficient of the identity of the group in PN. In the proof of Lemma 28.6 this was found to be $f = R \cap C : 1$. Hence $f' = n_{\alpha}/[(R \cap C : 1)(G : 1)]$, so that

$$(28.10) \qquad e = \left(n_{\alpha}/[(R \cap C : 1)(G : 1)]\right) \sum_{r \in R, c \in C} \theta(r)\phi(c)rc.$$

Note that since $e^2 = e$, this expression allows us to calculate n_{α}, if all other terms are known.

Let us return to the Young tableau [for its definition see Remarks, Section 23]. We will show that the conditions of Lemma 28.6 will be fulfilled when $\theta = 1$ on R and $\phi(c) = \pm 1$ according as $c \in C$ is an even or odd permutation. It will then follow immediately that the element e of (28.5) is a primitive idempotent.

First note that if a permutation p on the symbols 1, 2, ..., n is written concisely as

$$p = \begin{pmatrix} i \\ ip \end{pmatrix}, \qquad i = 1, 2, ..., n,$$

where ip is the symbol replacing i in the permutation p, then

$$sps^{-1} = \begin{pmatrix} is^{-1} \\ ips^{-1} \end{pmatrix}.$$

This means that if p is written as the product of disjoint cycles: $p = \cdots(i_1 i_2 \cdots i_k) \cdots$, then $sps^{-1} = \cdots (i_1 s^{-1} i_2 s^{-1} \cdots i_k s^{-1}) \cdots$.

Since R permutes only symbols in the rows, the symbols in any cycle of an element $r \in R$ are from a single row of the tableau. Now if $sRs^{-1} \cap C \neq 1$, then $s \notin CR$: thus if, say, $s = c_1 r_1$ and $c_1 r_1 r r_1^{-1} c_1^{-1} = c \neq 1$ then $r_1 r r_1^{-1} = c_1^{-1} c c_1$; but this is impossible since any cycle on the left side involves symbols from a single row whereas any on the right involve symbols from a single column. Similarly it can be shown that the converse holds. Suppose now that $sRs^{-1} \cap C \neq 1$: let $srs^{-1} = c \neq 1$. Let $(i_1 \cdots i_k)$ be one of the disjoint cycles of c. Then there is a disjoint cycle $(j_1 \cdots j_k)$ of r such that $j_1 s^{-1}, ..., j_k s^{-1} = i_1, ..., i_k$. Hence for the transposition $r' = (j_1 j_2)$ we have $s(j_1 j_2)s^{-1} = (i_1 i_2) = c'$. Since c' is an odd permutation $\phi(c') = -1$; but $\theta(r') = 1$ and so $\theta(r') \neq \phi(c')$. We have thus established that the conditions of Lemma 28.6 are satisfied when R and C are the subgroups of any tableau with $\theta = 1$ on R and $\phi = -1$ on the odd permutations of C. Thus each tableau yields a primitive idempotent e as given by (28.5). Then the minimal right ideal eA is a representation module for an irreducible representation. It will be shown in the next paragraph that the set of different shapes provide a complete system of idempotents from which all the irreducible representations can be constructed.

Let Σ_1, Σ_2 denote two tableaux of different shapes and let $e_1 = f_1 P_1 N_1$, $e_2 = f_2 P_2 N_2$ be their associated idempotents. Consider $P_1 N_1 s P_2 N_2$, $s \in S_n$. If $\forall r_2 = (ij) \in R_2$ there is no $c_1 \in C_1$ such that $sr_2 s^{-1} = c_1$, then no pair of symbols i, j from any row of Σ_2 can be taken by s^{-1} into symbols within a column of Σ_1 (otherwise for the transposition $(ij) \in R_2$ we would have $s(ij)s^{-1} = (is^{-1}js^{-1}) \in C_1$). Now this implies that the rows of Σ_1 have the same length as those of Σ_2 so that both have the same shape, contrary to assumption. This proves that $\exists r_2 = (ij)$ such that $sr_2s^{-1} = (is^{-1}js^{-1}) = c_1 \in C_1$. Now

$$P_1N_1sP_2N_2 = P_1N_1sr_2s^{-1}sP_2N_2 = P_1N_1c_1sP_2N_2$$
$$= \phi^{-1}(c_1)P_1N_1sP_2N_2 .$$

Hence $(1 - \phi^{-1}(c_1))P_1N_1sP_2N_2 = 0$ and since c_1 is a transposition $\phi^{-1}(c_1) \neq 1$ so that $P_1N_1sP_2N_2 = 0$. This holds $\forall s \in S_n$ and hence $e_1Ae_2 = 0$. Since e_1 and e_2 are primitive idempotents this is possible only if they belong to different blocks. This shows that the tableaux of different shapes yield primitive idempotents from distinct blocks and are thus associated with distinct irreducible representations. Now there is a one-to-one correspondence between shapes and classes of elements of S_n, which associates with each shape the class of permutations whose disjoint cycles have the lengths of its rows. For example the shape

$$\begin{matrix} \cdot & \cdot & \cdot \\ \cdot & \cdot & \\ \cdot & \cdot & \end{matrix}$$

corresponds to all those permutations which, when decomposed into disjoint cycles, consist of one 3-cycle and two 2-cycles. Recalling that two permutations are conjugate if and only if their cycle structure is the same, we see that there are many shapes as classes of conjugate elements. Since there are as many irreducible representations as conjugate classes (Theorem 18.1) the tableaux provide us with a complete system of primitive idempotents from which to construct all irreducible representations.

29. Some Examples of Group Representations

1. Cyclic Groups

Let $G = \langle g : g^n = 1\rangle$. If ϵ is a primitive nth root of unity then $\sigma_j(g) = \epsilon^j$, $j = 1, \ldots, n$ gives n distinct representations of G of degree one. Since G is abelian of order n this is the complete est of irreducible representations.

2. Abelian Groups

Let G be an abelian group of order n. By the fundamental theorem of abelian groups G is the direct product of cyclic groups: $G = C_1 \times C_2 \times \cdots \times C_l$, and for $i = 1, \ldots, l-1$, $n_i \mid n_{i+1}$, where n_i is the order of C_i. Of course $n = \prod_{i=1}^{l} n_i$. For $i = 1, \ldots, l$ let c_i generate C_i and let ϵ_i be a primitive n_ith root of unity. Then $\sigma_{j_1 \cdots j_l}$, $j_k = 1, 2, \ldots, n_k$, defined on the generators C_i by

$$\sigma_{j_1 \cdots j_l}(c_i) = \epsilon_i^{j_i}$$

gives all n irreducible representations of G, and all are of degree one.

3. The Symmetric Group S_n

We have seen in Section 28 that the Young tableaux give a routine method for the construction of all irreducible representations. In practice the work is tedious and for cases where n is small various *ad hoc* procedures can usually be employed. [See Example 1, p. 108.] However, an irreducible representation of degree $n-1$ is readily obtainable for we have:

(29.1) **Lemma.** *The natural representation of a doubly transitive permutation group is the sum of the* 1*-representation and an irreducible representation.*

A permutation group G is *doubly* transitive if any ordered pair of distinct symbols is taken into an arbitrary ordered pair of distinct symbols by some element of G. A representation μ of G is a natural representation if $\mu(g)$ permutes a basis of the corresponding representation module in the same way that g permutes the symbols (not elements) of G.

Proof. Let M with basis $\{m_1, ..., m_k\}$ be a representation module for the natural representation. Then if

$$g = \binom{i}{ig}, \qquad i = 1, ..., k,$$

$$m_i\mu(g) = m_i g = m_{ig}.$$

Let

$$M \sim a_1 M_1 \oplus \cdots \oplus a_l M_l$$

where $a_i M_i$ stands for the direct sum of a_i isomorphic irreducible submodules of M, and $M_i \not\simeq M_j$ if $i \neq j$. Hence by (11.5) and (11.9) since $\text{Hom}(M_i, M_j) = 0$ if $i \neq j$:

(29.2) $$i(M, M) = a_1^2 + \cdots + a_l^2.$$

We want to calculate $i(M, M)$. Let $\tau \in \text{Hom}(M, M)$; then by definition,

$$(m_i\tau)g = (m_i g)\tau = m_{ig}\tau, \qquad \forall g \in G.$$

Now if $m_i\tau = \sum_{j=1}^{k} f_i^j m_j$, $f_i^j \in F$, then we get

$$\sum_j f_i^j m_{jg} = \sum_j f_{ig}^j m_j = \sum_j f_{ig}^{jg} m_{jg},$$

the last step since jg runs through the same symbols as j.

$$\therefore \quad \forall i, j, g : f_i^j = f_{ig}^{jg}.$$

But by the double transitivity if $i \neq j$, $\exists g$ such that $ig = 1$, $jg = 2$.

$$\therefore \quad f_i^j = f_1^2, \qquad i \neq j,$$

and

$$f_i^i = f_{ig}^{ig} = f_1^1, \qquad \forall i.$$

$$\therefore \quad m_i\tau = f_1^1 m_i + f_1^2 \sum_{j \neq i} m_j.$$

If we now define τ_1 and τ_2 as the operator homomorphisms

$$m_i\tau_1 = m_i \,, \qquad \forall i,$$

$$m_i\tau_2 = \sum_{j \neq i} m_j \,,$$

then

$$\tau = f_1^1\tau_1 + f_1^2\tau_2 \,.$$

Thus Hom(M, M) is of rank 2, or $i(M, M) = 2$.

Equation (29.2) implies that there are only two $a_i \neq 0$ and that they are both one. Suppose $a_1 = a_2 = 1$. Then $M = M_1 \oplus M_2$, M_1, M_2 irreducible. Now if $m = \Sigma_{i=1}^{l} m_i$, then $mg = m$, $\forall g \in G$, so that $Fm = M_1$, say, is an invariant submodule of dimension 1. This proves the lemma.

Example 1. *The symmetric group* S_4. The number of conjugate classes is 5 and so there are 5 irreducible representations. The commutator subgroup is A_4, the alternating group consisting of all even permutations (Section 22, Exercise 1). Thus the number of linear characters is $S_4 : A_4 = 2$. They are, of course, the 1-representation and the representation which is -1 on odd permutations. By Lemma 29.1 there is a representation of degree 3. We proceed to find it.

Let $\{m_1, ..., m_4\}$ be an ordered basis of the module M which gives the natural representation. Take a new basis: $m_1' = m_1 + m_2 + m_3 + m_4$, $m_j' = m_j - m_1$, $j = 2, 3, 4$. Then

$$m_1'(12) = m_1' \,,$$

$$m_2'(12) = (m_2 - m_1)(12) = m_1 - m_2 = -m_2' \,,$$

$$m_3'(12) = (m_3 - m_1)(12) = m_3 - m_2 = (m_3 - m_1) - (m_2 - m_1) = m_3' - m_2' \,,$$

$$m_4'(12) = (m_4 - m_1)(12) = m_4 - m_2 = (m_4 - m_1) - (m_2 - m_1) = m_4' - m_2' \,.$$

Thus

$$\mu(12) = \begin{pmatrix} 1 & 0 & 0 & 0 \\ 0 & -1 & 0 & 0 \\ 0 & -1 & 1 & 0 \\ 0 & -1 & 0 & 1 \end{pmatrix}.$$

Similarly

$$\mu(234) = \begin{pmatrix} 1 & 0 & 0 & 0 \\ 0 & 0 & 1 & 0 \\ 0 & 0 & 0 & 1 \\ 0 & 1 & 0 & 0 \end{pmatrix}.$$

Now $\mu \sim 1 \oplus \tau$ by Lemma 29.1 so that we have the irreducible representation

$$\tau(12) = \begin{pmatrix} -1 & 0 & 0 \\ -1 & 1 & 0 \\ -1 & 0 & 1 \end{pmatrix}, \qquad \tau(234) = \begin{pmatrix} 0 & 1 & 0 \\ 0 & 0 & 1 \\ 1 & 0 & 0 \end{pmatrix}.$$

Since (234) and (12) generate S_4 (Coxeter [9], p. 7) τ can be found on the remaining elements.

We have now found 3 representations of degrees $n_1 = n_2 = 1$, $n_3 = 3$. Since the order of the group is 24,

$$24 = 1^2 + 1^2 + 3^2 + n_4^2 + n_5^2.$$

Thus $n_4 = 2$, $n_5 = 3$. This could also have been seen from the tableaux: the pairs of representations of equal degree correspond to the tableaux obtained by interchanging rows and columns. The unpaired representation of degree 2 arises from the symmetric tableau

```
. .
. .
```

Clearly if ϕ is the representation which is -1 on odd permutations and τ is any irreducible representation then $\tau' = \phi\tau$ is another irreducible representation. Accordingly

$$\tau'(12) = \begin{pmatrix} 1 & 0 & 0 \\ 1 & -1 & 0 \\ 1 & 0 & -1 \end{pmatrix},$$

$\tau'(234) = \tau(234)$ is the other irreducible representation of degree 3.

Finally we remark that S_4 has the Klein 4-group K_4 as a normal subgroup: $K_4 = \{1, (12)(34), (13)(24), (14)(23)\}$. Since $S_4/K_4 \cong S_3$, the irreducible representations of S_3 provide irreducible representations of S_4 [obvious: see also p. 93, part (2) of Lemma 26.1]. Now the irreducible representation of degree 2 derived from the natural representation of S_3 gives the representation of degree 2 of S_4.

Example 2. *The dihedral group* D_m. D_m has $2m$ elements and is the group of rotations of a regular polygon of m sides. It is generated by a, b with the defining relations: $a^2 = b^m = 1$, $a^{-1}ba = b^{-1}$. The element b corresponds to a rotation of the polygon through an angle of $2\pi/m$ about a perpendicular axis through its center; a corresponds to a 180° rotation about an axis of symmetry in its plane.

The commutator group $D'_m = \{1, b, ..., b^m\}$ if m is odd and is $\{1, b^2, ..., b^m\}$ if m is even. Thus $D_m : D'_m = 2$, or 4 and so there are 2 or 4 linear characters according as m is odd or even. In either case the cyclic subgroup N generated by b is of index 2 in D_m and so it is a normal subgroup. Then D_m/N is cyclic, of order 2, and its two representations of degree one are representations of D_m. If m is odd these are the only irreducible representations of degree one.

Case 1: $m = 2l + 1$. There are $l + 2$ conjugate classes: $\{1\}$, $\{ab^\beta : \beta = 1, ..., m\}$, $\{b^\gamma, b^{-\gamma}\}$, $\gamma = 1, ..., l$. Thus there are $l + 2$ irreducible representations of which two are of degree one, and l of degrees $n_i \geqslant 2$, $i = 1, ..., l$. Hence: $2(2l + 1) = 2 + \Sigma_{i=1}^{l} n_i^2 \geqslant 2 + 2^2 l$, showing that all $n_i = 2$. We have therefore for this case two irreducible representation of degree 1 and l of degree 2. Since $D_m/N \cong \{1, a\}$ the representations of degree one are obviously: $\sigma_1(a) = \sigma_1(b) = 1$; $\sigma_2(a) = -1$, $\sigma_2(b) = 1$. The remaining $\sigma_i(x)$ are 2×2 matrices which satisfy

$$(\sigma_i(b))^m = (\sigma_i(a))^2 = I_2 \quad \text{and} \quad \sigma_i(a)^{-1}\sigma_i(b)\sigma_i(a) = \sigma_i(b)^{-1}.$$

Remarking that

$$\begin{pmatrix} 0 & 1 \\ 1 & 0 \end{pmatrix}$$

is of order 2 and transforms a diagonal matrix by interchanging its elements, we can write down the $\sigma_i(x)$ by inspection. Thus:

$$(29.3)\quad \sigma_i(b) = \begin{pmatrix} \epsilon^i & 0 \\ 0 & \epsilon^{-i} \end{pmatrix}, \qquad \sigma_i(a) = \begin{pmatrix} 0 & 1 \\ 1 & 0 \end{pmatrix}, \qquad i = 1, 2, ..., l,$$

where ϵ is a primitive mth root of unity, satisfy the preceding relations and are surely the irreducible representations of degree 2. This can easily be checked by showing that $(\chi^{\sigma_i}, \chi^{\sigma_i}) = 1$.

Case 2: $m = 2l$. We have remarked that there are 4 representations of the first degree. The conjugate classes are: $\{1\}$, $\{ab, ab^3, ...\}$, $\{ab^2, ab^4, ..., a\}$, $\{b^\alpha, b^{-\alpha}\}$, $\alpha = 1, ..., l$. Again, $2(2l) = 4 + \sum_{i=1}^{l-1} n_i^2 \geqslant 4(l)$, so that as before all $n_i = 2$. The relations (29.3) still give representations. However, for σ_l we find

$$\sigma_l(b) = \begin{pmatrix} -1 & 0 \\ 0 & -1 \end{pmatrix}, \qquad \sigma_l(a) = \begin{pmatrix} 0 & 1 \\ 1 & 0 \end{pmatrix}$$

which on transformation by

$$T = \begin{pmatrix} 1 & 1 \\ 1 & -1 \end{pmatrix}$$

splits to give the two new representations $\sigma_1'(a) = 1$, $\sigma_1'(b) = -1$; $\sigma_2'(a) = -1$, $\sigma_2'(b) = -1$. These, together with σ_1, σ_2 as in Case 1 are the four irreducible representations of the first degree. Then the other $l - 1$ representations of degree 2 must be irreducible and we have found all irreducible representations.

Example 3. The tetrahedral group A_4. This is the group of rotations of a regular tetrahedron. It is isomorphic to A_4, the alternating group on four symbols, and can be generated by the cycles $a = (12)(34)$, $b = (123)$. There are four classes: $C_1 = \{1\}$, $C_2 = \{(12)(34), (13)(24), (14)(23)\}$, $C_3 = \{(123), (142), (134), (243)\}$, $C_4 = \{(124), (132), (234), (143)\}$. The Klein 4-group $K_4 = C_1 \cup C_2$ is a normal subgroup. Since $A_4/K_4 \cong \{1, (123), (123)^2\}$ is cyclic of order 3 its three irreducible representations are of degree one and are representations of A_4. Then $12 =$

$1^2 + 1^2 + 1^2 + n_4^2$, so that $n_4 = 3$. Now A_4 is doubly transitive on its four symbols so that σ_4 can easily be obtained as in Example 1. We find

$$\sigma_4(a) = \begin{pmatrix} -1 & 0 & 0 \\ -1 & 0 & 1 \\ -1 & 1 & 0 \end{pmatrix} \quad \text{and} \quad \sigma_4(b) = \begin{pmatrix} -1 & 1 & 0 \\ -1 & 0 & 0 \\ -1 & 0 & 1 \end{pmatrix}.$$

Going back to the representations of degree one we have $\sigma_i(a) = 1$, $\sigma_i(b) = \epsilon$, ϵ any cube root of unity. We list these as: $\sigma_1(a) = \sigma_1(b) = 1$; $\sigma_2(a) = 1$, $\sigma_2(b) = \omega$; $\sigma_3(a) = 1$, $\sigma_3(b) = \bar{\omega}$. Here $\omega = (-1 + \sqrt{3}i)/2$. We have now given all irreducible representations of A_4 on its generators.

Example 4. The icosahedral group. This is the group of rotations of a regular icosahedron. It is of order 60 and it is isomorphic to A_5, the alternating group on five symbols. A_5 can be generated by the cycles $a = (123)$, $b = (12345)$.

Since A_n, $n \geqslant 5$, is simple and not abelian, the commutator subgroup $A_5' = A_5$ and therefore the one-representation is the only irreducible representation of degree 1. Again, A_5 is doubly transitive on its five symbols so that there is an irreducible representation of degree 4 which can readily be written down as in Example 1.

The classes are $C_1 = \{1\}$; $C_2 = \{(12)(34), ...\}$ containing all 15 double transpositions; $C_3 = \{(123), ...\}$ with all 20 3-cycles; and two classes of 5-cycles containing 12 elements each: $C_4 = \{(12345), ...\}$; $C_5 = \{(13524), ...\}$. See Table I where the conjugate 5-cycles in each class are similarly underlined.

Since the class number is 5, there are 5 irreducible representations and we have

$$60 = 1^2 + 4^2 + n_3^2 + n_4^2 + n_5^2 .$$

Taking $n_3 \leqslant n_4 \leqslant n_5$ the last equation yields the unique solutions $n_3 = n_4 = 3$, $n_5 = 5$.

The irreducible representation σ_5 of degree 5 can be obtained as the irreducible component, $\neq 1$-representation, of the natural representation of a doubly transitive permutation group on six symbols.

Let us first note that if R is a subgroup of a group G and if $S_i = Rx_i$, $i = 1, \ldots, k$, are cosets relative to R, where the x_i are coset representatives, then

$$\pi(g) = \begin{pmatrix} S_i \\ S_i g \end{pmatrix}$$

represents G as a permutation group on its k cosets as symbols. Now A_5 contains a dihedral group D_5 as a subgroup of index 6 and can be represented as a doubly transitive permutation group on the six cosets. Table I is a multiplication table of D_5 by the set of

TABLE I

	c^5	c	c^2	c^3	c^4	w
	1	(14325)	(13542)	(12453)	(15234)	(14)(25)
b	(12345)	(154)	(253)	(14352)	(13)(24)	(15423)
b^2	(13524)	(123)	(15)(34)	(254)	(14532)	(132)
b^3	(14253)	(13452)	(124)	(15)(23)	(354)	(345)
b^4	(15432)	(24)(35)	(14523)	(134)	(125)	(12435)
d	(25)(34)	(142)	(13245)	(12354)	(153)	(143)
db	(12)(35)	(15243)	(234)	(145)	(13254)	(15324)
db^2	(13)(45)	(12534)	(152)	(243)	(14235)	(13425)
db^3	(14)(23)	(135)	(12543)	(15342)	(245)	(235)
db^4	(15)(24)	(23)(45)	(14)(35)	(13)(25)	(12)(34)	(12)(45)

$d^2 = b^5 = 1; \quad dbd = b^4$

coset representatives shown in the first line. Then each column gives the elements of a coset. The first five coset representatives themselves form a cyclic group of order 5. Let us represent

$b = (12345)$ as a permutation $\tilde{\pi}(b)$ of the coset representatives $1, c, c^2, c^3, c^2, w$. Now

$$\begin{aligned} 1b &= & &= b1 \\ cb &= (14325)(12345) &= (152) &= db^2c^2 \\ c^2b &= (13542)(12345) &= (143) &= dw \\ c^3b &= (12453)(12345) &= (13254) &= dbc^4 \\ c^4b &= (15234)(12345) &= (24)(35) &= b^4c \\ wb &= (14)(25)(12345) &= (15342) &= db^3c^3. \end{aligned}$$

Thus

$$\tilde{\pi}(b) = \begin{pmatrix} 1 & c & c^2 & c^3 & c^4 & w \\ 1 & c^2 & w & c^4 & c & c^3 \end{pmatrix}.$$

Since just one symbol is left unmoved $\chi^{\tilde{\pi}}(b) = 1$. But $\chi^{\sigma_5} = \chi^{\tilde{\pi}} - 1$ so that $\chi^{\sigma_5}(b) = 0$. In the same way we find

$$\chi^{\sigma_5}(1) = 5, \qquad \chi^{\sigma_5}(12)(34) = 1,$$
$$\chi^{\sigma_5}(123) = -1, \qquad \chi^{\sigma_5}(13524) = 0.$$

The representation σ_4 has already been dealt with. For example since

$$a = (123) = \begin{pmatrix} 1 & 2 & 3 & 4 & 5 \\ 2 & 3 & 1 & 4 & 5 \end{pmatrix},$$

if π is the natural representation, we get $\chi^{\pi}(a) = 2$ so that $\chi^{\sigma_4}(a) = \chi^{\pi}(a) - 1 = 1$. Thus we calculate the first line and the last two lines of the character table (Table II). The actual

TABLE II

	C_1	C_2	C_3	C_4	C_5
χ^1	1	1	1	1	1
χ^{σ_2}	3	-1	0	$(1 + \sqrt{5})/2$	$(1 - \sqrt{5})/2$
χ^{σ_3}	3	-1	0	$(1 - \sqrt{5})/2$	$(1 + \sqrt{5})/2$
χ^{σ_4}	4	0	1	-1	-1
χ^{σ_5}	5	1	-1	0	0

representations σ_4, σ_5 can be found in the same way as was the τ of Example 1. For instance, regarding the coset representatives as the basis of a module over the complex field F on which g acts like $\tilde{\pi}(g)$ [for example, $cb = c\tilde{\pi}(b) = c^2$], we get a representation module. Then the decomposition is accomplished by the new basis: $i' = (1 + c + c^2 + c^3 + c^4 + w)$; $c'^j = c^j - 1$, $j = 1, 2, ..., 4$; $w' = w - 1$. In this way we find

$$\sigma_5(a) = \begin{pmatrix} -1 & 0 & 0 & 0 & 1 \\ -1 & 0 & 0 & 1 & 0 \\ -1 & 1 & 0 & 0 & 0 \\ -1 & 0 & 1 & 0 & 0 \\ -1 & 0 & 0 & 0 & 0 \end{pmatrix}, \quad \sigma_5(b) = \begin{pmatrix} 0 & 1 & 0 & 0 & 0 \\ 0 & 0 & 0 & 0 & 1 \\ 0 & 0 & 0 & 1 & 0 \\ 1 & 0 & 0 & 0 & 0 \\ 0 & 0 & 1 & 0 & 0 \end{pmatrix}$$

on the generators $a = (123)$ and $b = (12345)$ of A_5.

Note that the $\tilde{\pi}(g)$ written as permutation matrices give the representation of A_5 induced by the one representation of D_5 [formula (23.6)]. The other linear representation of D_5 [Example 2] induces a representation τ of degree 6 which contains two irreducible representations* $\neq 1$. They can therefore only be the irreducible representations of degree 3. However, to obtain them it would be necessary to reduce τ. Instead, as an illustration, we will find these representations by the method of Lemma 28.6. The same method could be used to find all the representations in each of the examples considered so far. The conditions of Lemma 28.6 will be satisfied by the choice

$$\left.\begin{array}{lll} R = D_5, & \theta(b) = 1, & \theta(d) = \pm 1 \\ C = \{1, c, c^2, c^3, c^4\}, & \phi(c) = \eta, & \\ & \eta^5 = 1, & \eta \neq 1 \end{array}\right\} \text{ see Table I.}$$

This can be seen as follows: if $s \in CR$, since $R \cap C = 1$, then $sRs^{-1} \cap C = 1$ and the first condition is trivially satisfied because

* Check that $(\chi^\tau, \chi^\tau) = 2$. Alternatively, note that the proof of Lemma 29.1, except for the presence of the 1-representation, carries over for doubly transitive monomial representations (i.e., the basis elements are permuted doubly transitively, but also multiplied by a numerical factor). Briefly, $i(M, M) = 2$ remains valid.

$\theta(1) = \phi(1) = 1$. On the other hand, if $s \notin CR$, then $s^{-1} \notin RC$ and s^{-1} is in the last column of Table I. Thus $s^{-1} = r'w$, for some $r' \in R$. Now $sRs^{-1} = wr'^{-1}Rr'w = wRw \ni wbw = c^3$. Since $\theta(b) = 1$, $\phi(c^3) = \eta^3$, the second condition is satisfied as long as $\eta^3 \neq 1$, that is, η being a 5th root of unity, as long as $\eta \neq 1$.

Thus, putting

$$P = [(1 + b + b^2 + b^3 + b^4) \pm d(1 + b + b^2 + b^3 + b^4)]$$

and

$$N = [1 + c\eta + c^2\eta^2 + c^3\eta^3 + c^4\eta^4]$$

we have that $e = PN$ is a multiple of a primitive idempotent for each choice of sign in P and each choice of $\eta \neq 1$ in N. To distinguish the idempotents we will find the characters of the corresponding representations. We use the formula (p. 82 and Appendix)

$$\chi(g) = (n/(C(g):1)(R \cap C:1)) \sum_{rc \in C(g)} \theta(r)\phi(c)$$

where n, at present unknown, is the degree of the irreducible representation and the summation is over all r, c for which $rc \in C(g)$. This gives

$$\begin{aligned}
\chi(1) &= n, \\
\chi(12)(34) &= (n/15)[5(\pm 1) + (\pm\eta) + \eta + \eta^2 \\
&\qquad \pm \eta^2 + \eta^3 \pm \eta^3 + \eta^4 \pm \eta^4], \\
\chi(123) &= (n/20)[2\eta \pm 2\eta + 2\eta^2 \pm 2\eta^2 + 2\eta^3 \pm 2\eta^3 \\
&\qquad + 2\eta^4 \pm 2\eta^4], \\
\chi(12345) &= (n/12)[2 \pm 2\eta + 2\eta^2 + 2\eta^3 \pm 2\eta^4], \\
\chi(13542) &= (n/12)[2 + 2\eta \pm 2\eta^2 \pm 2\eta^3 + 2\eta^4].
\end{aligned}$$

Case 1: (+) taken in P.

$$\begin{aligned}
\chi(12)(34) &= (n/15)[3] = n/5, \\
\chi(123) &= (n/20)[-4] = -n/5, \\
\chi(12345) &= 0 = \chi(13524).
\end{aligned}$$

Since χ is irreducible,

$$1 = (\chi, \chi) = (1/60)(n^2 + 15n^2/25 + 20n^2/25) = n^2/25.$$

Therefore $n = 5$ and we have again the last line of Table II.

The four choices for η give idempotents PN_1, PN_2, PN_3, PN_4, which must then belong to the same block, that pertaining to the representation of degree 5. Actually the PN_i belong to the same minimal right ideal (see Case 2) and are part of a basis from which the representation may be obtained again.

Case 2: (−) *taken in P.*

$$\begin{aligned} \chi(12)(34) &= -n/3, \\ \chi(123) &= 0, \\ \chi(12345) &= (n/6)(1 - \eta + \eta^2 + \eta^3 - \eta^4), \\ \chi(13524) &= (n/6)[1 + \eta - \eta^2 - \eta^3 + \eta^4]. \end{aligned}$$

Here η may take any one of the values ω^k, $k = 1, 2, 3, 4$, where $\omega = e^{2\pi i/5} = \cos 72^\circ + i \sin 72^\circ$. Note that

$$1 + \omega + \omega^4 = (1 + \sqrt{5})/2, \; 1 + \omega^2 + \omega^3 = (1 - \sqrt{5})/2$$

and, generally, $1 + \eta + \eta^2 + \eta^3 + \eta^4 = 0$.

If $\eta = \omega$ or ω^4:

$$\chi(12345) = (n/6)(1 - \sqrt{5}), \qquad \chi(13524) = (n/6)(1 + \sqrt{5})$$

and if $\eta = \omega^2$ or ω^3:

$$\chi(12345) = (n/6)(1 + \sqrt{5}), \qquad \chi(13524) = (n/6)(1 - \sqrt{5}).$$

In both cases

$$\begin{aligned} 1 = (\chi, \chi) = (1/60)[n^2 + 15n^2/9 &+ 12(1 \pm \sqrt{5})^2 n^2/36 \\ &+ 12(1 \pm \sqrt{5})^2 n^2/36] = n^2/9. \end{aligned}$$

Therefore $n = 3$ and each pair of choices for η yields a distinct irreducible character of degree 3. These give the remaining two lines of Table II.

Let us turn now to the actual construction of these remaining representations. The choices ω and ω^4 for η give PN_1, PN_2 say, which, up to multiple, are primitive idempotents of the block pertaining to the character χ^{σ_3} (Table II). We assert that $PN_2 \in PN_1A$, A being the group algebra. This can be proved on general grounds but is easily seen for this case as follows: if $g = (23)(45) = db^4c$, then $gN_1g^{-1} = N_2$ because $gcg^{-1} = c^4$; then

$$PN_2 = PgN_1g^{-1} = Pdb^4(cN_1)g^{-1}$$
$$= \theta^{-1}(db^4)\phi^{-1}(c)PN_1g^{-1} \in PN_1A.$$

Thus PN_1, PN_2, denoted by e_1, e_2 belong to the same minimal right ideal of dimension 3 over F and, being independent, are elements of a basis. For convenience we shall take the remaining element of the basis to be $e_3 = PN_1w(1 + c + c^2 + c^3 + c^4)$, $w = (14)(25) = a^2c^4$ [Table I]. It is easily checked that the e_i are linearly independent: any relation $f_1e_1 + f_2e_2 + f_3e_3 = 0$, multiplied in succession by N_1, N_2 yields $f_1 = 0, f_2 = 0$, and so also $f_3 = 0$. Now

$$\begin{aligned} e_1c &= PN_1c = \phi_1^{-1}(c)PN_1 &&= \omega^{-1}e_1 \\ e_2c &= PN_2c = \phi_2^{-1}(c)PN_2 &&= \omega^{-4}e_2 \\ e_3c &= PN_1w(1 + c + c^2 + c^3 + c^4)c = &&\quad e_3 \end{aligned}$$

so that

$$\sigma_3(c) = \begin{pmatrix} \omega^{-1} & 0 & 0 \\ 0 & \omega & 0 \\ 0 & 0 & 1 \end{pmatrix}, \qquad c = (14325).$$

In general if x is any element of A_5, e_ix is a linear combination of e_1, e_2, e_3: $e_ix = f_{i1}e_1 + f_{i2}e_2 + f_{i3}e_3$, where the f_{ij} are complex numbers. For each i the f_{i1}, f_{i2}, f_{i3} may be found by solving the three equations obtained by equating the coefficients of 1, c, c^2 on both sides of the relation. Then $\sigma_3(x) = (f_{ij})$, a matrix of the third degree. Finally, to obtain the remaining representation σ_2 it is only necessary to replace ω by ω^2 in the matrices of σ_3.

EXERCISES

1. Prove that all irreducible characters of the symmetric group S_n lie in the rational field, and that the representations may be obtained in this field. [Look at the primitive idempotents.]

2. Show that a doubly transitive permutation group on n symbols cannot have fewer than $n^2 - 2n + 2$ elements.

3. Find all the irreducible representations of the quaternion group $G = \langle a, b : a^4 = 1, a^2 = b^2, a^{-1}ba = b^{-1} \rangle$.

4. Show that the subgroups $R = A_4$, $C = \{(12345)^k : k = 1, 2, \ldots, 5\}$ of A_5 with their linear characters satisfy the conditions of Lemma 28.6. Classify all the resulting primitive idempotents and give bases for the corresponding irreducible representations.

5. Use the expression (28.5) for the idempotent arising from the tableau

$$\begin{array}{|l|}\hline 123 \cdots n-1 \\ \hline n \\ \hline \end{array},$$

or its transpose, to show that S_n has an irreducible representation of degree $n - 1$.

6. Find an $n \times n$ matrix $\neq I_n$ which permutes with all elements of a doubly transitive group of permutation matrices. [Cf. $\tau_2 \in \mathrm{Hom}(M, M)$ in the proof of Lemma 29.1, for example.]

CHAPTER VI

Modular Representations

30. General Remarks

In Chapter III we dealt with certain complex-valued functions on a group, namely, the ordinary characters. Recall that if σ is a matrix representation of a group G which assigns to each $g \in G$ a matrix $\sigma(g)$ with coefficients in the field of complex numbers, then the ordinary character χ^σ on G is defined by $\chi^\sigma(g) =$ trace $\sigma(g)$. If σ is irreducible χ^σ is called an irreducible character. We saw that the irreducible characters on a group satisfy certain character relations and moreover that any character can be expressed uniquely in terms of them. Sections 26 and 27 showed the importance of the ordinary characters in answering questions about the group.

For further insight into the structure of the characters certain other complex-valued functions on a group, called *Brauer characters*, must be considered. These can be defined by means of the representation of the finite group in a modular field, that is, a field of finite characteristic p. The connection with the ordinary characters is established through the *integral representations*. For example, suppose that σ is a representation of G in the field P of rational numbers. It will presently be shown in a more general context that σ is equivalent to another representation τ with the property that $\forall g \in G$ the matrix $\tau(g)$ has coefficients in J, the ring of rational integers. Now τ is an example of an integral representation. If A is the group algebra of G over P and if $\mathfrak{O}$ is the set of all linear combinations: $c_1 g_1 + c_2 g_2 + \cdots + c_n g_n$ of group elements g_i over the rational integers c_i then it is easy to see that $\mathfrak{O}$ is a subring of A. $\mathfrak{O}$ is an example of an order in A. Moreover $\forall x \in \mathfrak{O}$, $\tau(x)$ is a matrix with integer coefficients.

If now each integer in the matrix $\tau(x)$ is replaced by its residue class modulo a fixed prime p, we obtain a matrix $\overline{\tau(x)}$ with coefficients in a finite field of characteristic p which may be taken as $\Pi = \{1, 2, ..., p - 1, 0\}$. Thus $\bar{\tau}(x) = \overline{\tau(x)}$ is a representation in Π and $\bar{\tau}$ is called the modular representation of $\mathfrak{O}$, and so of G, corresponding to the integral representation τ. Every integral representation gives rise to a modular representation in this way, but it should be noted that not every representation at characteristic p can be so obtained as a modular representation. It should also be understood that even if τ is irreducible, nonetheless $\bar{\tau}$ is in general reducible in Π or some extension field of Π. As a final note of caution we remark that trace $\bar{\tau}(g)$, which is an element in Π, though a character in our former sense, is *not* what we will define as a Brauer character. The Brauer, or modular character $\psi^{\bar{\tau}}$ will be a complex-valued function derived from $\bar{\tau}$ in a special way.

The example of an integral representation and its corresponding modular representation given in the preceding paragraph is really too narrow to fit the general case. In the first place it is not possible to have all the ordinary representations of a group G lying in the rational field P. The theory of Chapter III was developed for the case where the representing field F is of characteristic $p = 0$ or $p \nmid G:1$ and F is algebraically closed; specifically, F is the field of all algebraic numbers: numbers which are solutions of equations of the form

$$(30.1) \qquad x^r + c_1x^{r-1} + c_2x^{r-2} + \cdots + c_{r-1}x + c_r = 0$$

where the c_i are rational numbers; moreover, every such equation has its solutions in F.

For an integral representation we are again dealing with the case $p = 0$, but in general $\sigma(g)$ lies in $F - P$. However, since the group G is of finite order n the set of matrices $\{\sigma(g)\}$, for all $g \in G$ and all irreducible representations σ, is finite. If we now adjoin to P the finite set of coefficients of all these matrices we get a field B, in which all the representations lie, and $P \subset B \subset F$. Moreover, we know from the theory of field adjunction that B can be obtained by the adjunction of a single element $\theta \in F : B = P(\theta)$, and that the degree of B over P is finite: $[B : P] = m$. Also θ satisfies an irreducible equation of degree m having the form of (30.1).

In the general case now described, an integral representation τ has its matrices $\tau(g)$ with coefficients in the *integers of* B. These are the numbers of B which satisfy an equation, (30.1), in which the coefficients c_i are rational integers. It can be shown that these integers form a ring J_B and that B is the quotient field of J_B. Thus every $x \in B$ can be expressed as $x = a/b$, $a, b \in J_B$. So far the relations between J_B and B parallel those between J and P. Now we come to a second distinction between the present case and our former simple example: whereas J is a principal ideal ring, in general J_B is not. This means that each ideal $L \subset J$ has an element a such that $L = aJ$. On the other hand, an ideal $L_B \subset J_B$ may require two generators $a, b \in J_B$ so that $L_B = aJ_B + bJ_B$.

Now in our formulation it will be essential to work with principal ideal rings. To restore the situation we will embed B in a p-adic field E whose integers $\mathfrak{o}$ include J_B. In this way the local situation in J_B will be the same. However, $\mathfrak{o}$ is a principal ideal ring having E as its quotient field.

The next sections are devoted to a detailed exposition of this outline. The development of the theory of modular representations is mainly the work of R. Brauer. We follow here a treatment due to him. However, no account of the Block theory is given. For this we refer the reader to Brauer [5], Osima [17] as well as the text by Curtis and Reiner [10].

31. p-Regular Elements of a Finite Group

Before the Brauer characters can be defined it is necessary to introduce some new terms and to prove a number of propositions. This will be done here and in the next section.

(31.1) **Definition.** *Let G be a group of finite order $n = p^a h$ where p is a prime integer and $(p, h) = 1$. An element $g \in G$ is said to* *be p-regular if its order m is relatively prime to p; that is, m is the smallest positive integer for which $g^m = 1$ and $(m, p) = 1$.*

Note that if an element g is p-regular then all elements which are conjugate to g, and thus have the same order, are also p-regular. Hence we can consider p-regular classes of conjugate elements.

(31.2) **Lemma.** *Let g be an element of order $p^b k$ where k is relatively prime to p. Then there are two unique elements x and y in G such that x is p-regular, the order of y is a power of p, and $g = xy = yx$. x is called the p-regular factor of g.*

Proof. Since $(p^b, k) = 1$ there are integers α and β so that $\alpha p^b + \beta k = 1$. Then $g = g^{\alpha p^b + \beta k} = g^{\alpha p^b} g^{\beta k} = g^{\beta k} g^{\alpha p^b}$. Writing $x = g^{\alpha p^b}$ and $y = g^{\beta k}$ we have

$$g = xy = yx.$$

Moreover, since $x^k = g^{\alpha p^b k} = (g^{p^b k})^\alpha = 1$ and $y^{p^b} = (g^{p^b k})^\beta = 1$, the order of x divides k and so is relatively prime to p while the order of y divides p^b and so is a power of p. Thus x and y have the required properties. Now suppose that $g = x_1 y_1 = y_1 x_1$, and that x_1 is p-regular and the order of y_1 is a power of p. Let m be the order of x and m_1 that of x_1. Then $(m, p) = 1$, $(m_1, p) = 1$ and we have $(l, p) = 1$ where l is the least common multiple of m and m_1. Then noting that y_1 commutes with g and hence with $g^{\beta k} = y$ and similarly that x_1 commutes with x:

$$g = xy = x_1 y_1 \Rightarrow x x_1^{-1} = y^{-1} y_1 \Rightarrow (x x_1^{-1})^{p^s} = (y^{-1} y_1)^{p^s} = 1$$

for some suitable power p^s. Since $(x x_1^{-1})^l = 1$, if d is the order of $x x_1^{-1}$ we have $d \mid l$ and $d \mid p^s$. As l and p^s are relatively prime this is impossible unless $d = 1$. Hence $x x_1^{-1} = 1 = y^{-1} y_1 \Rightarrow x = x_1$ and $y = y_1$, and the uniqueness is established.

(31.3) **Lemma.** *If $g = xy$, where x is the p-regular factor of g, and if σ is any matrix representation of G in a field of characteristic p, then the matrices $\sigma(g)$ and $\sigma(x)$ have the same characteristic roots.*

Proof. We know from linear algebra that there is a matrix P such that

$$P^{-1}\sigma(g)P = \begin{pmatrix} \lambda_1 & & & \\ & \lambda_2 & & \\ & & \ddots & \\ & * & & \lambda_d \end{pmatrix}$$

and $\lambda_1, ..., \lambda_d$ are the characteristic roots of $\sigma(g)$. Now from Lemma 31.2 x and y are powers of g and the order of y is a power of p. Let $x = g^r$, $y = g^t$, $y^{p^s} = 1$. Then

$$P^{-1}\sigma(y^{p^s})P = \begin{pmatrix} 1 & & & \\ & 1 & & \\ & & \ddots & \\ & & & 1 \end{pmatrix} = P^{-1}\sigma(g^{tp^s})P = \begin{pmatrix} \lambda_1^{tp^s} & & \\ & \ddots & \\ & & \lambda_d^{tp^s} \end{pmatrix}$$

and we see that $(\lambda_i^t)^{p^s} = 1$. But the field is of characteristic p so that

$$0 = (\lambda_i^t)^{p^s} - 1 = (\lambda_i^t - 1)^{p^s} \Rightarrow \lambda_i^t = 1.$$

Thus

$$P^{-1}\sigma(y)P = P^{-1}\sigma(g^t)P = \begin{pmatrix} \lambda_1^t & & \\ & \ddots & \\ * & & \lambda_d^t \end{pmatrix} = \begin{pmatrix} 1 & & & \\ & 1 & & \\ & & \ddots & \\ & * & & 1 \end{pmatrix},$$

so that

$$\begin{aligned} P^{-1}\sigma(g)P &= P^{-1}\sigma(xy)P = P^{-1}\sigma(x)P \cdot P^{-1}\sigma(y)P \\ &= P^{-1}\sigma(g^r)P \cdot P^{-1}\sigma(y)P \\ &= \begin{pmatrix} \lambda_1^r & & & \\ & \lambda_2^r & & \\ & & \ddots & \\ & * & & \lambda_d^r \end{pmatrix} \cdot \begin{pmatrix} 1 & & & \\ & 1 & & \\ & & \ddots & \\ & * & & 1 \end{pmatrix} \end{aligned}$$

$$= \begin{pmatrix} \lambda_1^r & & \\ & \ddots & \\ * & & \lambda_d^r \end{pmatrix} = \begin{pmatrix} \lambda_1 & & \\ & \ddots & \\ * & & \lambda_d \end{pmatrix}$$

whence $\lambda_i^r = \lambda_i$. Since the λ_i^r are characteristic roots of $\sigma(g^r) = \sigma(x)$ and the λ_i are characteristic roots of $\sigma(g)$ we have the result.

32. Conditions for Two Representations to Have the Same Composition Factors

If two representations have the same composition factors it is clear that their characters coincide. If the field of the representations is of characteristic p and $p = 0$ or $p \nmid G : 1$ we have seen (Lemma 19.13) that the converse is true. Moreover, because of the complete reducibility (Corollary 16.5) which holds in this case, if the characters coincide, the representations are equivalent. In the case $p \mid G : 1$ this is no longer true. Two representations can here have the same character but different composition factors, or having the same composition factors, fail to be equivalent due to the lack of complete reducibility. [Thus if σ is indecomposable and

$$\sigma(g) = \begin{pmatrix} F_1(g) & \\ * & F_2(g) \end{pmatrix},$$

$F_i(g)$ irreducible matrices, then

$$\tau(g) = \begin{pmatrix} F_1(g) & \\ 0 & F_2(g) \end{pmatrix}$$

is a representation and has the same irreducible constituents yet $\sigma \nsim \tau$ by assumption.] For the general case we have:

(32.1) **Lemma.** *Let F be an algebraically closed field of characteristic p. Let $\rho_1, \rho_2, \ldots, \rho_s$ be the irreducible representations of an algebra A over F. If the representation σ contains each irreducible* *constituent ρ_i with multiplicity a_i and the representation τ contains ρ_i with multiplicity b_i then σ and τ have the same character if and only if $a_i \equiv b_i \pmod{p}$.*

Proof. (1) If $a_i \equiv b_i \pmod{p}$, then in F, $\forall x \in A$,

$$\chi^{\sigma}(x) = \sum_{i=1}^{s} a_i \chi^{\rho_i}(x) = \sum_{i=1}^{s} b_i \chi^{\rho_i}(x) = \chi^{\tau}(x).$$

(2) Assume that $\forall x$, $\chi^{\sigma}(x) = \chi^{\tau}(x)$. It follows from the Frobenius-Schur theorem (Theorem 10.10) that for any fixed element $x' \in A$ and any j, $1 \leqslant j \leqslant s$, we can find an element $x^* \in A$ such that $\rho_i(x^*) = 0$, $i \neq j$, and $\rho_j(x^*) = \rho_j(x')$. Now $\rho_j(A)$ is an irreducible algebra of matrices and by the Burnside theorem (Theorem 13.1) it is a complete matrix algebra. Hence $\rho_j(A)$ contains a matrix E with 1 in its upper left-hand corner and zeros elsewhere. If we take x' so that $\rho_j(x') = E$ then

$$\chi^{\rho_i}(x^*) = 0, \qquad i \neq j, \qquad \chi^{\rho_j}(x^*) = \chi^{\rho_j}(x') = 1 \in F.$$

Then

$$\chi^{\sigma}(x^*) = \sum_{i=1}^{s} a_i \chi^{\rho_i}(x^*) = a_j \cdot 1 = \chi^{\tau}(x^*) = \sum_{i=1}^{s} b_i \chi^{\rho_i}(x^*) = b_j \cdot 1.$$

Therefore

$$a_j \cdot 1 = b_j \cdot 1$$

in the field F of characteristic p. Since a_j and b_j are rational integers we have $a_j \equiv b_j \pmod{p}$.

Although two group representations with different composition factors can have the same character the next theorem shows that their characteristic roots cannot be the same for each group element.

(32.2) **Theorem.** *Let F be an algebraically closed field of characteristic p. Let σ and τ be two matrix representations in F of a finite group G. Then σ and τ have the same composition factors if and only if $\sigma(g)$ and $\tau(g)$ have the same characteristic roots for every $g \in G$.*

Proof. Let $\rho_1, \rho_2, \ldots, \rho_s$ be the irreducible representation of G. Suppose that

$$(32.3) \qquad \sigma(g) \sim \begin{pmatrix} \ddots & & & & \\ & \rho_i(g) & & & \\ & & \ddots & & \\ & * & & \rho_i(g) & \\ & & & & \ddots \end{pmatrix} \quad \Big\} \; a_i \text{ times}$$

and

$$\tau(g) \sim \begin{pmatrix} \ddots & & & & \\ & \rho_i(g) & & & \\ & & \ddots & & \\ & * & & \rho_i(g) & \\ & & & & \ddots \end{pmatrix} \quad \Big\} \; b_i \text{ times}$$

where $a_i \geqslant 0$ and $b_i \geqslant 0$. The characteristic roots of $\sigma(g)$ are the characteristic roots of each ρ_i, each repeated a_i times and, changing a_i to b_i, the same statement holds for $\tau(g)$. If the composition factors are the same we have $a_i = b_i$ and it is clear that the characteristic roots of $\sigma(g)$ and $\tau(g)$ are identical.

Now suppose that the characteristic roots of $\sigma(g)$ and $\tau(g)$ are the same. Then the degree of σ = the degree of $\tau = m$, say. Since the trace of a matrix is the sum of its characteristic roots we have

$$\chi^\sigma(g) = \chi^\tau(g), \qquad \forall g \in G,$$

and thus $\chi^\sigma(x) = \chi^\tau(x)$ for every x in the group algebra A of G over F. By Lemma 32.1

$$(32.4) \qquad a_i \equiv b_i \qquad (\text{mod}\, p).$$

We make an induction argument on the degree of σ (= the degree of τ as remarked), supposing the theorem true for all pairs of representations of equal degree $<m$. The case $m = 1$ is obvious.

Let us replace σ and τ by the representations σ' and τ' derived from σ and τ, respectively, by replacing the $*$ by 0 in (32.3). This does not alter the a_i and b_i but we may now write the direct sums

$$\sigma' = \sum_{i=1}^{s} a_i\rho_i \,, \qquad \tau' = \sum_{i=1}^{s} b_i\rho_i \,.$$

Case 1. There is a j such that $a_j \neq 0$ and $b_j \neq 0$. Then remove one ρ_j from each side to get the representations

$$\sigma'' = \sum_{i=1}^{s} \hat{a}_i\rho_i \,, \qquad \tau'' = \sum_{i=1}^{s} \hat{b}_i\rho_i$$

in which $\hat{a}_i = a_i$, $\hat{b}_i = b_i$, $i \neq j$, and $\hat{a}_j = a_j - 1$, $\hat{b}_j = b_j - 1$. Since σ'' and τ'' are of degree $<m$ the inductional assumption gives $\hat{a}_i = \hat{b}_i$ and so $a_i = b_i$ for $i = 1, 2, \ldots, s$ and hence the composition factors of σ and τ coincide.

Case 2. $\forall j$, $a_j = 0$, *or* $b_j = 0$. Let $a_j \neq 0$, then $b_j = 0$. But from (32.4), $a_j \equiv b_j \pmod{p}$. Therefore $a_j = pa'_j$. Similarly, if $b_j \neq 0$, $a_j = 0$, and we have $b_j = pb'_j$. Thus

$$\sigma' = p\left(\sum_{j=1}^{s} a'_j\rho_j\right) \qquad \text{and} \qquad \tau' = p\left(\sum_{j=1}^{s} b'_j\rho_j\right).$$

Then $\sigma'' = \Sigma_{j=1}^{s} a'_j\rho_j$ and $\tau'' = \Sigma_{j=1}^{s} b'_j\rho_j$ are representations of degree $<m$ and so $a'_j = b'_j$. Thus $a_j = a'_j p = b'_j p = b_j$ and again the composition factors of σ and τ are identical.

Since these are the only cases possible the theorem is proved.

33. The Brauer Characters

Let G be a finite group of order $n = p^m h$, where p is a prime integer and $(p, h) = 1$. Let σ be a matrix representation of G in an algebraically closed field F of characteristic p.

We know from Lemma 31.3 that the characteristic roots of the matrix $\sigma(g)$ are the same as those of $\sigma(x)$, where x is the p-regular factor of g. Since $x^h = 1$, the characteristics roots of $\sigma(x)$, and hence of $\sigma(g)$, are hth roots of unity. Let ϵ be a primitive hth root of unity, so that the other hth roots of unity are suitable powers ϵ^t of ϵ. Now let ξ be a primitive hth root of unity in the field of complex numbers. Then the Brauer character ψ^σ of σ is defined (on p-regular elements) as follows: $\psi^\sigma(g) = \Sigma\, \xi^t$, where the sum is over the exponents t of every characteristic root ϵ^t of $\sigma(g)$. The achievement of the next result is the motivation for considering these modified characters.

(33.1) **Lemma.** *Two modular representations have the same Brauer character if and only if they have the same irreducible constituents.*

Proof. (1) If σ and τ have the same irreducible constituents then $\forall g$, $\sigma(g)$ and $\tau(g)$ have the same set of characteristic roots ϵ^t. Thus $\psi^\sigma(g) = \psi^\tau(g)$, $\forall g \in G$.

(2) Suppose now that the Brauer characters are the same. Let $\sigma(g)$ have the characteristic roots $\epsilon^{a_1}, \epsilon^{a_2}, \ldots, \epsilon^{a_f}$ while those of $\tau(g)$ are $\epsilon^{b_1}, \epsilon^{b_2}, \ldots, \epsilon^{b_l}$. Then, taking the ith power of each term in each set, we get the characteristic roots of $\sigma(g^i)$ and $\tau(g^i)$. But then

$$(*) \qquad \xi^{ia_1} + \xi^{ia_2} + \cdots + \xi^{ia_f} = \xi^{ib_1} + \xi^{ib_2} + \cdots + \xi^{ib_l}.$$

Now consider the representations

$$\sigma'(g^i) = \begin{pmatrix} \xi^{ia_1} & & \\ & \ddots & \\ & & \xi^{ia_f} \end{pmatrix}, \qquad \tau'(g^i) = \begin{pmatrix} \xi^{ib_1} & & \\ & \ddots & \\ & & \xi^{ib_l} \end{pmatrix}$$

of the cyclic group $H = \{g^i\}$.

The relation $(*)$ implies that $\chi^{\sigma'}(g) = \chi^{\tau'}(g)$, $\forall g \in H$. But these are ordinary characters in the complex field, and our previous results (Lemma 19.13) show that their irreducible constituents are the same. Thus, for $i = 1$, the sequences $\{\xi^{a_1}, ..., \xi^{a_f}\}$ and $\{\xi^{b_1}, ..., \xi^{b_l}\}$ are equal up to a permutation. Therefore $f = l$ and one set is merely a rearrangement of the other. This proves the result.

34. Integral Representations

We are going to consider integral representations in a field E which satisfies certain special conditions. It will be seen later that the general case can be reduced to a consideration of this particular case. Accordingly, let E be a field having a subset $\mathfrak{o}$ with the following properties:

(1) $\mathfrak{o}$ is a ring with identity.

(2) $\mathfrak{o}$ is a principal ideal ring. This means that every ideal $\mathfrak{a}$ in $\mathfrak{o}$ can be written $\mathfrak{a} = a\mathfrak{o}$, where a is a particular element of $\mathfrak{a}$. In other words, every ideal consists of multiples of a particular element in it.

(3) E is the quotient field of $\mathfrak{o}$: every element $x \in E$ can be written $x = a/b$ where $a, b \in \mathfrak{o}$ and $b \neq 0$. We will refer to $\mathfrak{o}$ as integers of the field E.

Presently E will be taken to be a p-adic field in which the properties (1) to (3) always hold. Besides, there will then be a topological structure which allows the valuable notions of limits and convergent sequences.

Now a matrix representation σ in E of a group G is an integral representation if $\forall g \in G$ the matrix $\sigma(g)$ has all its coefficients in $\mathfrak{o}$, the integers of E as now called. Since we want to consider more generally integral representations of an algebra A over E it is necessary to consider the notion of an $\mathfrak{o}$-module and especially that of an order $\mathfrak{O}$ in A. We have already considered general modules in Definition 3.1. The $\mathfrak{o}$-module M' considered now is a subset of an E-module M (vector space M) with the properties: (1) $m' \in M'$, $m'' \in M' \Rightarrow m' + m'' \in M'$. (2) $\forall \alpha \in \mathfrak{o}$, $\forall m' \in M' : \alpha m' \in M'$. [We do *not* require $\beta m' \in M'$ if $\beta \in E$.] In the present case we have:

(34.1) **Lemma.** *Let* $\mathfrak{o}$ *be a principal ideal ring and let* E *be its quotient field. Let* M *be an* E*-module having a subset* M' *which is an* $\mathfrak{o}$*-module. If* M' *has a finite* $\mathfrak{o}$*-basis* $m'_1, m'_2, \ldots, m'_l$ *and if* $s \leqslant l$ *is the maximal number of the* m'_i *which are linearly independent over* E*, then* M' *has an* $\mathfrak{o}$*-basis* $m''_1, \ldots, m''_s$*, and these are linearly independent over* E*.*

Proof. If $s = l$ there is nothing to prove, for the m'_i are already the m''_i. Assume $s < l$. We may suppose that $\{m'_1, \ldots, m'_s\}$ is the maximal set of the m'_i which are linearly independent over E. Then

$$(34.2) \qquad m'_j = \sum_{i=1}^{s} t^i_j m'_i, \qquad t^i_j \in E, \qquad j = s+1, \ldots, l.$$

Since E is the quotient field of $\mathfrak{o}$ we have $t^i_j = a^i_j/b^i_j$, a^i_j, $b^i_j \in \mathfrak{o}$, and $b^i_j \neq 0$. What we do now is, in effect, to find the least common multiple b of all the b^i_j in the system (34.2). [Thus, let $\mathfrak{B}$ be the intersection of all the ideals $b^i_j\mathfrak{o}$:

$$\mathfrak{B} = \bigcap_{i,j} (b^i_j\mathfrak{o}).$$

Now $\mathfrak{B}$ is an ideal and is not the zero ideal since $0 \neq \prod_{i,j} b^i_j \in \mathfrak{B}$. Since $\mathfrak{o}$ is a principal ideal ring, the ideal $\mathfrak{B}$ is generated by a single element $b : \mathfrak{B} = b\mathfrak{o}$. But $b \in$ each $(b^i_j\mathfrak{o})$, therefore $b = b^i_j c^i_j$, $c^i_j \in \mathfrak{o}$.] Then $t^i_j = a^i_j/b^i_j = a^i_j c^i_j / b^i_j c^i_j = a^i_j c^i_j / b$. Writing $d^i_j = a^i_j c^i_j$ and $m''_i = (1/b)m'_i$, $i = 1, 2, \ldots, s$, (34.2) becomes

$$(34.3) \qquad m'_j = \sum_{i=1}^{s} d^i_j m''_i, \qquad j = s+1, \ldots, l.$$

The m''_i are linearly independent over E because so are the m'_i. Since the m'_i are an $\mathfrak{o}$-basis of M', $\forall m' \in M'$, we have

$$m' = \sum_{i=1}^{s} u^i m'_i + \sum_{j=s+1}^{l} v^j m'_j, \qquad u^i, v^j \in \mathfrak{o},$$

and so using (34.3)

$$m' = \sum_{i=1}^{s} u^i \left(b \frac{1}{b} m_i'\right) + \sum_{j=s+1}^{l} v^j \left(\sum_{i=1}^{s} d_j^i m_i''\right)$$

$$= \sum_{i=1}^{s} \left(u^i b + \sum_{j=s+1}^{l} v^j d_j^i\right) m_i'' .$$

Since the coefficients in parentheses are in $\mathfrak{o}$ we see that the m_i'', $i = 1, 2, \ldots, s$, are an $\mathfrak{o}$-basis of M' which are linearly independent over E. Thus the lemma is proved.

Now let A be an algebra, with an identity element, of rank n over the field E. As before, E is the quotient field of a principal ideal ring $\mathfrak{o}$. We consider E as embedded in A by identifying the identity elements.

(34.4) **Definition.** *A subset $\mathfrak{O}$ of A is an* ***order*** *in A if:*

(1) *$\mathfrak{O}$ is a subring of A, containing the identity element.*

(2) *$\mathfrak{O}$ is a finitely generated $\mathfrak{o}$-module.*

(3) *$\mathfrak{O}$ contains an E-basis of A.*

For example, if $A(G)$ is the group algebra of a finite group G over E, then

$$\mathfrak{O} = \{c_1 g_1 + \cdots + c_n g_n : g_i \in G, \quad c_i \in \mathfrak{o}\}$$

is easily seen to be an order in $A(G)$. If σ is a matrix representation of $A(G)$ such that $\forall g \in G$, $\sigma(g)$ has its coefficients in $\mathfrak{o}$ (the integers of E), then it is clear that $\forall x \in \mathfrak{O}$, $\sigma(x)$ also has its coefficients in $\mathfrak{o}$. Thus we have a motivation for:

(34.5) **Definition.** *An integral representation of an algebra A over E is a representation σ for which there is an order $\mathfrak{O}$ in A such that for each x in $\mathfrak{O}$ the matrix $\sigma(x)$ has its coefficients in $\mathfrak{o}$.*

(34.6) **Theorem.** *Every representation of an algebra A over E is similar to an integral representation.*

Proof. Let μ be a representation of A afforded by an E-A module M which has a basis $m_1, m_2, \ldots, m_d$ over E. Suppose that $w_1, w_2, \ldots, w_n$ is an $\mathfrak{o}$-basis of an order $\mathfrak{O}$ in A. Let M' be the $\mathfrak{o}$-module generated by the set $S = \{m_i w_j\}$, $i = 1, 2, \ldots, d$, $j = 1, 2, \ldots, n$. Clearly $M'\mathfrak{O} \subset M'$. Now S has d linearly independent elements over E (S generates M' and M' contains the E-basis $m_1, m_2, \ldots, m_d$ of A, since $1 \in \mathfrak{O}$). Then by Lemma 34.1 M' has an $\mathfrak{o}$-basis $m_1', \ldots, m_d'$ which are linearly independent over E, and so are an E basis of M. Hence if w is any element in $\mathfrak{O}$,

$$m_i' w = \sum_{j=1}^{d} c_{ij} m_j', \qquad c_{ij} \in \mathfrak{o}.$$

Thus, using the basis $\{m_i'\}$ we get a representation σ such that if $w \in \mathfrak{O}$, $\sigma(w) = (c_{ij})$, a matrix with coefficients in $\mathfrak{o}$, so that σ is an integral representation. Since $m_1', \ldots, m_d'$ are also an E basis of M, the representations σ and μ are similar.

(34.7) **Definition.** *Two integral representations σ and τ are* ***integrally equivalent*** *if there exists a matrix T such that T and T^{-1} have all their coefficients in $\mathfrak{o}$ and*

$$T^{-1}\sigma(w)T = \tau(w), \qquad \forall w \in \mathfrak{O}.$$

Integral equivalence implies similarity. That the converse is false is shown by the following example:

Let G be the cyclic group of order 2: $G = \{a\}$, $a^2 = 1$. Take E as P, the field of rational numbers so that $\mathfrak{o}$ is the ring J of ordinary integers. If

$$\sigma(a) = \begin{pmatrix} 1 & 0 \\ 0 & -1 \end{pmatrix} \quad \text{and} \quad \tau(a) = \begin{pmatrix} 1 & 1 \\ 0 & -1 \end{pmatrix}$$

it can be seen that σ and τ are similar (equivalent) but not integrally equivalent. Thus, let

$$\begin{pmatrix} x & y \\ u & v \end{pmatrix}\begin{pmatrix} 1 & 0 \\ 0 & -1 \end{pmatrix} = \begin{pmatrix} 1 & 1 \\ 0 & -1 \end{pmatrix}\begin{pmatrix} x & y \\ u & v \end{pmatrix}.$$

If we multiply out and equate coefficients we find $u = 0$, $v = -2y$. Hence the most general T is

$$T = \begin{pmatrix} x & y \\ 0 & -2y \end{pmatrix}.$$

But then

$$T^{-1} = \begin{pmatrix} 1/x & +1/2x \\ 0 & -1/2y \end{pmatrix}$$

and T, T^{-1} cannot both have integral coefficients.

EXERCISE

Let ρ be the regular representation of an algebra A of finite rank n over the field E. If $\mathfrak{A} = \{a \mid a \in A, \rho(a)$ with coefficients in $\mathfrak{o}\}$, show that $\mathfrak{A}$ is an order in A. Conversely, if $\mathfrak{O}$ is an order in A and a basis of $\mathfrak{O}$ is used for the regular representation ρ of A, show that $\mathfrak{O} = \mathfrak{A}$.

35. Ordinary and Modular Representations of Algebras

1. Arithmetic in an Algebra

We continue to consider the special case in which E is the quotient field of a principal ideal integral domain $\mathfrak{o}$ and A is an algebra, with identity, over E. Let us recall the following facts about a principal ideal domain $\mathfrak{o}$.

(I) Every element $a \in \mathfrak{o}$ can be written uniquely as a product of prime elements of $\mathfrak{o}$: $a = p_1^{\alpha_1} \cdots p_k^{\alpha_\kappa}$. [Consider the ideal $a\mathfrak{o}$, if it is not maximal, we have the ideal chain $a\mathfrak{o} = \mathfrak{a}_1 \subset \mathfrak{a}_2 \subset \cdots$, none of them the unit ideal. Then $\vartheta = \bigcup \mathfrak{a}_i$ is an ideal not the unit ideal, and since $\mathfrak{o}$ is a principal ideal ring $\exists p \in \vartheta$ such that $\vartheta = p\mathfrak{o}$. But also since $p \in \vartheta$ there is a first index j for which $p \in \mathfrak{a}_j$. Hence $\vartheta = p\mathfrak{o} \subset \mathfrak{a}_j$ so that the members of the chain are identical after $\mathfrak{a}_j$, which is therefore maximal. Moreover, $a \in p\mathfrak{o}$

and so $a = pc$. Continuing in this way with c, and beyond if necessary, we arrive at a factorization of a. Note that the process must terminate, since, for example $a\mathfrak{o} \subset c\mathfrak{o}$ so that it gives rise to an ascending chain. Finally, if $p = xy$, then $p\mathfrak{o} \subseteq x\mathfrak{o}$. If $p\mathfrak{o} \subset y\mathfrak{o}$, then by the maximality of $p\mathfrak{o}$, $y\mathfrak{o} = \mathfrak{o}$. Thus $\exists s$ such that $ys = 1$. Hence y is unit in $\mathfrak{o}$ and so p and x are associates (like 3 and -3 for integers). On the other hand, if $p\mathfrak{o} = y\mathfrak{o}$, then $y = pt = xyt \Rightarrow y(1 - xt) = 0$ and since there are no divisors of zero in $\mathfrak{o}$, $xt = 1$, whence x is a unit and so p and y are associates.]

(II) If p is a prime, the prime ideal $p\mathfrak{o}$ is maximal in $\mathfrak{o}$ (this was seen in (I)) and the residue class ring $\bar{\mathfrak{o}} = \mathfrak{o}/p\mathfrak{o}$ is a field. [If $a \in \mathfrak{o}$ and $a \notin p\mathfrak{o}$ then the ideal $a\mathfrak{o} + p\mathfrak{o}$, written $(a\mathfrak{o}, p\mathfrak{o})$, properly contains $p\mathfrak{o}$ so that $(a\mathfrak{o}, p\mathfrak{o}) = \mathfrak{o}$. Then $\exists x \in \mathfrak{o}$, $y \in \mathfrak{o}$, such that $ax + py = 1$, therefore $ax \equiv 1 \pmod{p}$. This means that modulo p, every $a \neq 0$ has an inverse.]

Now let $\mathfrak{O}$ be an order in A and let p be a fixed prime in $\mathfrak{o}$.

(35.1) **Definition.** *A representation $\bar{\sigma}$ of $\mathfrak{O}$ in the field $\bar{\mathfrak{o}} = \mathfrak{o}/p\mathfrak{o}$ is called a* ***modular representation*** *of $\mathfrak{O}$.*

It should be noted that a modular representation as thus defined need not be a representation in a modular field. The latter refers to a representation in a field of characteristic $p \neq 0$ whereas $\bar{\mathfrak{o}}$ could have characteristic zero. For example if $\mathfrak{o} = P[x]$, P the rational field, x an indeterminate, then $p = x$ is a prime; but $\bar{\mathfrak{o}} = P[x]/p\mathfrak{o} \cong P$, of characteristic 0.

It is clear that $p\mathfrak{O}$ is an ideal in $\mathfrak{O}$ and, since p as a field element commutes with all elements of A, even a two-sided ideal. Moreover since we have identified the identity elements of E and A and since $1 \in \mathfrak{o}$ and $1 \in \mathfrak{O}$ therefore $p\mathfrak{O} \supset p\mathfrak{o}$. Intuitively, we may think of the elements of $\mathfrak{O}$ as "integers" of the algebra A. For this reason arithmetical properties, that is, properties concerning the ideal structure, of $\mathfrak{O}$ are called arithmetic in the algebra A. This subject will not be treated to any extent here. Let the elements $w_1, w_2, \ldots, w_n$, which are linearly independent over E, be an $\mathfrak{o}$-basis of $\mathfrak{O}$. Then $\mathfrak{O}$ consists of elements of the form $c_1w_1 + \cdots + c_nw_n$, $c_i \in \mathfrak{o}$, and $p\mathfrak{O}$ are those elements, and only those, for which each c_i is a multiple of p. Then the residue class ring $\bar{\mathfrak{O}} = \mathfrak{O}/p\mathfrak{O}$ consists of the elements $\bar{c}_1\bar{w}_1 + \cdots + \bar{c}_n\bar{w}_n$, where the bars indicate a residue class* modulo $p\mathfrak{O}$. Any relation $\bar{c}_1\bar{w}_1 + \cdots + \bar{c}_n\bar{w}_n = 0$ implies that $c_1w_1 + \cdots + c_nw_n \in p\mathfrak{O}$ so that all c_i are multiples of p and $\bar{c}_i = 0$. Thus we see that the $\bar{w}_i$ are linearly independent over the field $\bar{\mathfrak{o}}$ and that $\bar{\mathfrak{O}}$ is an algebra of rank n over $\bar{\mathfrak{o}}$.

Let $\bar{\sigma}$ be a modular representation of $\mathfrak{O}$. Since $\bar{\sigma}(p\mathfrak{O}) = \bar{\sigma}(p)\bar{\sigma}(\mathfrak{O}) = (pI)\bar{\sigma}(\mathfrak{O}) \equiv 0$ modulo p, the ideal $p\mathfrak{O}$ lies in the kernel of $\bar{\sigma}$. Because of this, $\bar{\sigma}$ induces a representation $\bar{\bar{\sigma}}$ of the $\bar{\mathfrak{o}}$-algebra $\bar{\mathfrak{O}}$ defined as follows:

$$\bar{\bar{\sigma}}(\bar{a}) = \bar{\sigma}(a), \qquad a \in \mathfrak{O}.$$

Conversely if $\bar{\bar{\sigma}}$ is a representation of $\bar{\mathfrak{O}}$ the same definition provides a modular representation of $\mathfrak{O}$. This shows that the modular representations of $\mathfrak{O}$ are essentially the same as the $\bar{\mathfrak{o}}$ representations of the $\bar{\mathfrak{o}}$-algebra $\bar{\mathfrak{O}}$.

EXERCISE

If a and b are elements of $\mathfrak{o}$ show that $a \equiv b$ modulo $p\mathfrak{o}$ is equivalent to $a \equiv b$ (modulo $p\mathfrak{O}$).

2. Connection with Integral Representations

Let σ be a matrix representation of an algebra A over E. If σ is an integral representation with respect to the order $\mathfrak{O}$ in A, then $\forall x \in \mathfrak{O}$, $\sigma(x)$ is a matrix with coefficients in $\mathfrak{o}$. Replacing each coefficient in $\sigma(x)$ by its residue class modulo $p\mathfrak{o}$ we get a matrix which we shall denote by $\overline{\sigma(x)}$. Now the representation $\bar{\sigma}$ defined by

$$\bar{\sigma}(x) = \overline{\sigma(x)}, \qquad x \in \mathfrak{O},$$

* See exercise at the close of this section.

is a modular representation of $\mathfrak{O}$. Every integral representation σ gives rise to a corresponding modular representation $\bar{\sigma}$ in this way. However, we shall see that not every representation in a modular field $\bar{\mathfrak{o}}$ can be derived from an integral representation.

So far the ring $\mathfrak{o}$ is not specific enough for connecting ordinary representations and modular representations in a given modular field. As we have seen $\mathfrak{o}$ may even have characteristic zero. We will work with the case where $\mathfrak{o}$ is a local ring, that is, a ring having exactly one maximal ideal $\mathfrak{p}$. Such a ring can be obtained as the ring of integers of a p-adic field. For this reason we are going to consider algebras over p-adic fields. The p-adic theory that will be needed is developed in the next section.

36. p-Adic Fields

In this section we will develop all the p-adic theory that we shall need. The only outside appeal will be to Hensel's lemma, and a result on the equivalence of the norms of a finite-dimensional vector space.

1. General Definition and Properties

A complete p-adic field E is a field with the properties:

(i) E has characteristic 0.

(ii) There is defined on E an exponential valuation. This means that for every $x \in E$ there is a real number $l(x)$ such that if $a, b \in E$,

$$l(ab) = l(a) + l(b),$$
$$l(a + b) \geqslant \text{minimum}\{l(a), l(b)\},$$
$$l(0) = \infty.$$

By convention: $\infty + \infty = \infty = \infty + \alpha$, $\forall$ real number α.

(iii) The valuation is discrete; that is, the value set $\{l(a) | \, a \in E\}$ has no limit point.

(iv) There exists a rational integer m such that $l(m) \neq 0$. Here we identify m with $m \cdot 1 = 1 + 1 + \cdots + 1$, $1 \in E$.

(v) E is complete: Cauchy's convergence theorem holds for sequences in E.

Conditions (i)–(iv) define a p-adic field.

(36.1) **Theorem.** *If E is a p-adic field the set of elements $\mathfrak{o} = \{x : x \in E, \quad l(x) \geqslant 0\}$ is a principal ideal ring, with identity, containing exactly one maximal ideal $\mathfrak{p} = \{x : l(x) > 0\}$. All ideals of $\mathfrak{o}$ are powers of $\mathfrak{p}$. The residue class ring $\mathfrak{o}/\mathfrak{p}$ is a field of characteristic p for some unique rational prime p.*

Proof. Since $l(1) = l(1 \times 1) = l(1) + l(1)$, therefore $l(1) = 0$ and $1 \in \mathfrak{o}$. Then $l(1) = 0 = l(-1 \times -1) = l(-1) + l(-1)$ and so $l(-1) = 0$. Also $l(-x) = l(-1 \cdot x) = l(-1) + l(x) = l(x)$. Hence $x \in \mathfrak{o} \Rightarrow -x \in \mathfrak{o}$.

Now if $x, y \in \mathfrak{o}$, we have $l(x + y) \geqslant \text{minimum}\{l(x), l(y)\} \geqslant 0$ and $l(xy) = l(x) + l(y) \geqslant 0$, so that $x + y$ and $xy \in \mathfrak{o}$. This proves that $\mathfrak{o}$ is a ring with identity.

Since $0 = l(1) = l(xx^{-1}) = l(x) + l(x^{-1})$ we see that $l(x^{-1}) = -l(x)$. Thus x and x^{-1} can both belong to $\mathfrak{o}$ if and only if $l(x) = 0$, a condition which therefore characterizes the units of $\mathfrak{o}$.

Since $\mathfrak{p} = \{x : l(x) > 0\}$, it is clear that $\mathfrak{p} \subset \mathfrak{o}$. If $x \in \mathfrak{p}$, $l(-x) = l(x) > 0$, and so $-x \in \mathfrak{p}$. If $x \in \mathfrak{p}$ and $y \in \mathfrak{o}$, $l(xy) = l(x) + l(y) > 0$, so that $xy \in \mathfrak{p}$. Moreover, if x, $y \in \mathfrak{p}$, then $l(x + y) \geqslant \min\{l(x), l(y)\} > 0$ so that $x + y \in \mathfrak{p}$.

$$\therefore \quad \mathfrak{p} \text{ is an ideal in } \mathfrak{o}.$$

Since $\mathfrak{p}$ consists of all nonunits of $\mathfrak{o}$ it is maximal and contains all other ideals. Therefore $\mathfrak{p}$ is the unique maximal ideal in $\mathfrak{o}$.

The set $\{l(x), x \in \mathfrak{p}\}$ is a discrete set of positive real numbers. Thus it has a smallest number $l(s) > 0$. Then, if $x \in \mathfrak{p}$, $l(x) \geqslant l(s)$, so that $l(x) - l(s) = l(x) + l(s^{-1}) = l(xs^{-1}) \geqslant 0$. Therefore $xs^{-1} \in \mathfrak{o}$ so that $x \in s\mathfrak{o}$ and we have that $\mathfrak{p} = s\mathfrak{o}$.

Similarly, for any ideal $\mathfrak{a}$ there is an element a such that $\mathfrak{a} = a\mathfrak{o}$. Therefore $\mathfrak{o}$ is a principal ideal ring.

Since $\mathfrak{a} \subset \mathfrak{p}$, $l(a) \geqslant l(s)$ and there is a unique positive integer $n \geqslant 1$ such that $(n + 1)l(s) > l(a) \geqslant nl(s)$. Therefore

$$l(s) > l(a) - nl(s) = l(a) - l(s^n) = l(a/s^n) \geqslant 0.$$

But $l(s)$ is the smallest positive value on $\mathfrak{p}$ so that $l(a/s^n) = 0$. Therefore

$$a/s^n = t, \qquad \text{a unit of } \mathfrak{o}.$$

Therefore

$$\mathfrak{a} = a\mathfrak{o} = s^n t\mathfrak{o} = s^n t t^{-1}\mathfrak{o} = s^n\mathfrak{o} = (s\mathfrak{o})^n = \mathfrak{p}^n.$$

If $b \in \mathfrak{o} - \mathfrak{p}$, $l(b) = 0 = -l(b) = l(b^{-1})$, so that $b^{-1} \in \mathfrak{o} - \mathfrak{p}$; that is, if $b \not\equiv 0 \pmod{\mathfrak{p}}$, there exists $b^{-1} \in \mathfrak{o}$ such that $bb^{-1} = 1 \bmod \mathfrak{p}$). Therefore the residue class ring $\mathfrak{o}/\mathfrak{p}$ is a field.

Finally, since there is an integer m such that $l(m) \neq 0$,

$$0 \neq l(m) = l(1 + 1 + \cdots + 1) \geqslant \min\{l(1), l(1), ...\} = l(1) = 0.$$

Therefore $l(m) > 0$ and so $m \in \mathfrak{p}$. Let $m = p_1^{\alpha_1} p_2^{\alpha_2} \cdots p_j^{\alpha_j}$ where the p_i are rational primes. Then

$$l(m) = l(p_1^{\alpha_1} \cdots p_j^{\alpha_j}) = \alpha_1 l(p_1) + \cdots + \alpha_j l(p_j) > 0.$$

Hence there is a prime $p_i = p$ such that $l(p) > 0$, which means that $p \in \mathfrak{p}$. Therefore $\mathfrak{o}/\mathfrak{p}$ is of characteristic p, a fact which implies the uniqueness of p. It follows then that $l(q) = 0$ for any other prime q. The proof of the theorem is now complete.

2. Ordinary Valuation. Metrical Properties

Property (ii) of Section 1 permits us to define a valuation V on E which possesses the fundamental properties of a metric or ordinary absolute value.

Let c be a fixed real number such that $0 < c < 1$. For $a \in E$ define

$$V(a) = c^{l(a)}, \qquad V(0) = 0 \qquad (c^\infty = 0).$$

From property (ii) of Section 1 we get:

(1) $V(a) > 0, \quad V(a) = 0$ if and only if $a = 0$.

(2) $V(ab) = V(a)V(b)$.

(3) $V(a + b) \leqslant \text{maximum}\{V(a), V(b)\} \leqslant V(a) + V(b)$.

This valuation with these three properties could have been used instead of the exponential valuation in defining a p-adic field. Note that $V(1) = V(-1) = 1$, $V(-a) = V(a)$, and from (3)

$$| V(a) - V(b)| \leqslant V(a - b) \leqslant V(a) + V(b).$$

The set of values $\{V(a) : a \in E\}$ has zero as its only limit point by (iii) of Section 1.

We now define limits and Cauchy sequences exactly as in Analysis: an element α of E is a p-adic limit of a sequence $a_1, a_2, \ldots, a_n, \ldots$ if for any $\epsilon > 0$, $V(\alpha - a_n) < \epsilon$ for all sufficiently large n. The sequence is a Cauchy sequence if for any $\epsilon > 0$, $V(a_m - a_n) < \epsilon$ for all sufficiently large m, n.

The requirement (v) of Section 1 concerning completeness means that every Cauchy sequence in E must have a p-adic limit in E.

3. Completion of a p-Adic Field

(36.2) **Theorem.** *Every p-adic field F can be extended to a complete p-adic field E.*

Proof. The construction required here is an exact parallel of that for the completion of the rational numbers to the reals. Denote a sequence $a_1, a_2, \ldots, a_n, \ldots$ of elements in F by $[a_n]$. Let $R = \{[a_n]\}$ be the set of all Cauchy sequences in F. If $[a_n]$, $[b_n] \in R$ we define their sum and product as follows:

(1) $[a_n] + [b_n] = [a_n + b_n]$,

(2) $[a_n][b_n] = [a_n b_n]$.

If $[a_n]$ is a Cauchy sequence, given any $\epsilon > 0$, there is an integer M such that $V(a_n - a_M) < \epsilon$, whenever $n > M$.

Then

$$V(a_n) = V(a_n - a_M + a_M) \leqslant V(a_n - a_M) + V(a_M) \leqslant \epsilon + V(a_M),$$

if $n > M$. Therefore

$$V(a_n) \leqslant \max\{V(a_1), \ldots, V(a_{M-1}), \quad \epsilon + V(a_M)\} \qquad \text{for all } n,$$

showing that the values of a Cauchy sequence are bounded. Since

$$(3) \quad V((a_m + b_m) - (a_n + b_n)) \leqslant V(a_m - a_n) + V(b_m - b_n) < \epsilon$$

for m, n sufficiently large, and also

$$\begin{aligned}(4) \quad V(a_m b_m - a_n b_n) &= V(a_m(b_m - b_n) + b_n(a_m - a_n)) \\ &\leqslant V(a_m)V(b_m - b_n) + V(b_n)V(a_m - a_n) \\ &\leqslant k_1 V(b_m - b_n) + k_2 V(a_m - a_n) < \epsilon, \\ &\qquad\qquad m, n \text{ large enough,}\end{aligned}$$

we see that with the sum and product (1) and (2) R is a ring. Note that $(1, 1, \ldots, 1, \ldots)$ is the identity and $(0, 0, \ldots, 0, \ldots)$ is the zero element of R.

Now call a (necessarily Cauchy) sequence $[a_n]$ a *null sequence* if $\lim_{n\to\infty} V(a_n) = 0$. Let N be the set of all null sequences. If $[a_n]$, $[b_n] \in N$ and $[c_n] \in R$,

$$V(a_n + b_n) \leqslant V(a_n) + V(b_n) \to 0,$$

and

$$V(a_n c_n) = V(a_n)V(c_n) \to 0$$

as $n \to \infty$. This shows that N is an ideal in R.

If $[a_n] \in R$ but $[a_n] \notin N$, then there exists M, ϵ such that $V(a_n) > \epsilon > 0$ when $n > M$. Thus $\forall a_n > 0$, if $n > M$, write $\angle a_n]$ for the (necessarily Cauchy) sequence: $(0, 0, \ldots, 0, 1/a_{M+1}, \ldots)$. Then

$$\begin{aligned}[a_n]\angle a_n] &= (0, 0, \ldots, 0, 1, 1, 1, \ldots, 1, \ldots) \\ &= (1, 1, \ldots) - (1, 1, \ldots, 1, 0, 0, \ldots) \\ &= (1, 1, \ldots) \qquad (\text{mod } N).\end{aligned}$$

Hence every sequence which is in R but not in N has an inverse in R/N. Therefore the residue class ring $E = R/N$ is a field. The sequences $(a, a, a, ...)$ form a subfield of $E = R/N$ which is isomorphic to F, and which we identify with F. In this way $E \supset F$ and so an extension has been constructed.

We wish to extend the valuation V to E. If $\alpha \in E$, α is defined, mod N, by a Cauchy sequence $[a_n]$. Then the sequence $[V(a_n)]$ is a Cauchy sequence of real numbers and has a limit in the real numbers. We define

$$\tilde{V}(\alpha) = \lim_{n-\infty} V(a_n).$$

The definition is independent of the representative sequence chosen, for if $[a_n] \equiv [b_n]$ mod N, $\lim V(a_n) = \lim V(b_n)$. It must be checked that $\tilde{V}$ possesses the properties (1)–(3) of Section 2. This is easy if it is borne in mind that $\alpha \in E$ means that α is represented by a Cauchy sequence $[a_n]$. It is left as an exercise to the reader. If $\alpha \in F$, $\alpha \sim (a, a, ...)$ and $\tilde{V}(\alpha) = \lim V(a) = V(a)$ so that $\tilde{V}$ is an extension of V.

Let $\alpha \in E$ be represented by $[a_n]$. Each a_n, identified with $(a_n, a_n, ...)$, is also in E. We show that $\lim a_n = \alpha$. Given $\epsilon > 0$, choose M so that $V(a_n - a_m) < \epsilon$, if $m, n > M$. Then $\tilde{V}(\alpha - a_m) = \lim_{n\to\infty} V(a_n - a_m) \leqslant \epsilon$, $m > M$. Since $\epsilon > 0$ is arbitrary $\tilde{V}(\alpha - a_m) \to 0$ and $\alpha = \lim a_m$, by the definition of limit.

Finally it must be shown that E is complete. Let $[\alpha_n]$, $\alpha_n \in E$, be a Cauchy sequence. Let $\alpha_n \sim [a_i^n]$. Then $\lim_{i\to\infty} a_i^n = \alpha_n$, so that i_n can be chosen to ensure that

$$\tilde{V}(\alpha_n - a_i^n) < 1/n, \qquad i > i_n .$$

The sequence $a_{i_1}^1, a_{i_2}^2, ..., a_{i_n}^n, ...$ is a Cauchy sequence for

$$V(a_{i_m}^m - a_{i_n}^n) \leqslant V(a_{i_m}^m - \alpha_m) + V(\alpha_n - a_{i_n}^n) + V(\alpha_n - \alpha_m)$$

$$\leqslant 1/m + 1/n + \epsilon,$$

$$V(a_{i_m}^m - a_{i_n}^n) \to 0, \qquad \text{as} \qquad m, n \to \infty.$$

Hence $[a^n_{i_n}] \sim \alpha \in E$ and so $\lim_{n\to\infty} a^n_{i_n} = \alpha$. But

$$V(\alpha - \alpha_n) \leqslant V(\alpha - a^n_{i_n}) + V(a^n_{i_n} - \alpha_n) < \epsilon,$$

for n large enough. Therefore

$$\lim \alpha_n = \alpha,$$

so that E is complete.

4. p-Adic Valuation of the Rational Field

Let E be a p-adic field. Let $l(x)$ be its exponential valuation. Since E has characteristic zero it contains the rational numbers P as a subfield. Then the restriction of l to P will provide an exponential valuation of the rational field.

We know from the proof of Theorem 36.1 (final line) that there is a unique rational prime p such that $l(p) > 0$ whereas $l(q) = 0$ for any other prime q. Thus for any integer m not containing p as a factor we have $l(m) = 0$. Put $l(p) = k > 0$.

Now for any rational number a we may write

$$a = (s/t)p^d \tag{1}$$

where s and t are integers prime to p and d is an integer. Hence

$$l(a) = l(s) - l(t) + dl(p) = dk. \tag{2}$$

Thus the values of l on P are integral multiples of a fixed positive constant.

Conversely, for any prime p, an arbitrary rational number can be written as in (1) and a function $l(x)$ can be defined on P by (2). It is easy to show that this $l(x)$ is an exponential valuation on P. Thus every p-adic valuation of the rational field is given by (2).

5. Extension of the p-Adic Valuation to Algebraic Number Fields

Let P be the rational field. For a fixed prime p and a fixed positive constant k define the exponential valuation $l(x)$ on P as

$$l(a) = dk, \tag{1}$$

where $a = (s/t)p^d$; s, t prime to p. Let Ω_p denote the p-adic completion of P [Section 3]. By Theorem 36.1 $\mathfrak{o} = \{x : x \in \Omega_p, l(x) \geqslant 0\}$ is a principal ideal ring and has a unique maximal ideal $\mathfrak{p} = \{x : l(x) > 0\}$ which contains p, since $l(p) = k > 0$. $\mathfrak{o}$ is called the ring of p-adic integers.

Now let E be an extension field, of finite degree n, over Ω_p. Thus there is a basis $e_1, e_2, \dots, e_n$ of elements in E such that any element $\alpha \in E$ can be expressed uniquely in the form

$$\alpha = \alpha_1 e_1 + \alpha_2 e_2 + \cdots + \alpha_n e_n, \qquad \alpha_i \in \Omega_p.$$

We may regard E as its own representation space and, using this basis, find its regular representation ρ:

$$e_i \rho(\alpha) = e_i \alpha = \sum_{j=1}^{n} \alpha_{ij} e_j, \qquad i = 1, 2, \dots, n.$$

This gives rise to a matrix (α_{ij}) and we write $\rho(\alpha) = (\alpha_{ij})$. The norm of α, denoted by $N(\alpha)$, is defined to be

$$N(\alpha) = \text{determinant of } \rho(\alpha).$$

Finally we define

$$\tilde{l}(\alpha) = (1/n) l(N(\alpha)). \tag{36.3}$$

(36.4) **Theorem.** *$\tilde{l}(\alpha)$ is an exponential valuation on E and is an extension of the valuation l* on *Ω_p. With respect to this valuation E is a complete p-adic field.*

The proof of the theorem requires a number of results the first of which, a special case of Hensel's lemma, is stated without proof. (See van der Waerden [22], Vol. I, p. 248).

The elements of $\mathfrak{o} = \{x : x \in \Omega_p, l(x) \geqslant 0\}$ are p-adic integers. If $a, b \in \mathfrak{o}$ we say that a divides b, if $l(b/a) \geqslant 0$, i.e., if $b/a \in \mathfrak{o}$.

A polynomial

$$a_0 x^r + a_1 x^{r-1} + \cdots + a_r, \qquad a_i \in \mathfrak{o},$$

is called primitive if there is no integer $b \neq 1$, dividing all a_i. Any polynomial becomes primitive if all a_i are divided by a_ν, where $l(a_\nu) = \text{minimum}\{l(a_i)\}$.

(36.5) **Reducibility Criterion** (Hensel's lemma). *Let $f(x)$ be a primitive polynomial with integral coefficients in Ω_p. Let $g_0(x)$ and $h_0(x)$ be two polynomials with integral coefficients in Ω_p which satisfy*

$$f(x) \equiv g_0(x)h_0(x) \qquad (\text{mod } \mathfrak{p}).$$

Then there exist two polynomials $g(x)$, $h(x)$ with integral coefficients in Ω_p for which

$$\begin{aligned} f(x) &= g(x)h(x) \\ g(x) &= g_0(x) \qquad (\text{mod } \mathfrak{p}) \\ h(x) &= h_0(x) \qquad (\text{mod } \mathfrak{p}) \end{aligned}$$

provided that $g_0(x)$ and $h_0(x)$ are relatively prime modulo $\mathfrak{p}$. It is, moreover, possible to determine $g(x)$ and $h(x)$ so that the degree of $g(x)$ is equal to the degree of $g_0(x)$ modulo $\mathfrak{p}$.

We deduce:

(36.6) **Lemma.** *If $x^m + a_1x^{m-1} + \cdots + a_{m-1}x + a_m = 0$ is an irreducible equation with coefficients in Ω_p and if a_m is a p-adic integer ($a_m \in \mathfrak{o}$) so are all a_i.*

Proof. (a) If all $l(a_i) \geqslant 0$ there is nothing to prove, since all a_i are then integers.

(b) Suppose $\exists l(a_i) < 0$. Let a_ν be such that $l(a_\nu) = \min\{l(a_i)\}$ then $l(a_\nu) < 0$. By dividing the equation by a_ν we get the irreducible primitive polynomial

$$(*) \qquad f(x) = b_0x^m + b_1x^{m-1} + \cdots + b_m = 0$$

where $b_0 = 1/a_\nu$, $b_m = a_m/a_\nu$. Since $l(1) = 0$, and since $l(a_m) \geqslant 0$ by hypothesis, we have

$$l(b_0) = -l(a_\nu) > 0; \qquad l(b_m) \doteq l(a_m) - l(a_\nu) > 0.$$

As $f(x)$ is primitive not all b_i are divisible by $\mathfrak{p}$ so that $\exists\, b_j$ such that $l(b_j) = 0$. Moreover, by the last relations $j \neq 0, m$. Let b_μ be the first b_i in $(*)$ for which $l(b_i) = 0$. Then $l(b_t) > 0$ for $t = \mu - 1$, $\mu - 2, \ldots, 0$ and $\mu \geqslant 1$. We now have

$$f(x) \equiv 1 \cdot (b_\mu x^{m-\mu} + b_{\mu+1}x^{m-\mu-1} + \cdots + b_m) \qquad (\text{mod } \mathfrak{p})$$

which is impossible, since by the reducibility criterion we should then have $f(x)$ decomposable into a factor of degree μ and one of degree $m - \mu$.

Thus (a) is the only case and the lemma is proved.

(36.7) **Definition.** *An element γ of the extension field E is* ***integral with respect to the ring*** $\mathfrak{o}$ *of p-adic integers, or is* ***a p-adic algebraic integer,*** *if it satisfies an irreducible equation*

$$x^m + a_1x^{m-1} + \cdots + a_m = 0$$

having unity for its leading coefficient and having all other coefficients in $\mathfrak{o}$.

Note that the set of all algebraic p-adic integers is a ring. This follows exactly as in Lemma 24.3 and Corollary 24.5.

(36.8) **Lemma.** *An element $\gamma \in E$ is a p-adic algebraic integer if and only if $\tilde{l}(\gamma) \geqslant 0$.*

Proof. Recall that $\tilde{l}(\gamma) = (1/n)l(N(\gamma))$, where the norm of γ is $N(\gamma) = \det \rho(\gamma)$ and ρ is the regular representation of E using a basis of E over Ω_p. Let

$$(*) \qquad f(x) = x^m + a_1x^{m-1} + \cdots + a_m = 0$$

be the irreducible equation satisfied by γ. Then

$$(\rho(\gamma))^m + a_1(\rho(\gamma))^{m-1} + \cdots + a_m = 0$$

and $(*)$ is the minimum equation of the matrix $\rho(\gamma)$. Since the minimum equation is irreducible we know from matrix theory that the characteristic equation of $\rho(\gamma)$ is a power of its minimum equation:

$$\therefore \quad \det(xI - \rho(\gamma)) = (f(x))^s, \qquad s > 0.$$

Putting $x = 0$ we get $N(\gamma) = \pm a_m^s$, so that $\tilde{l}(\gamma) = (s/n)l(a_m)$. Therefore $\tilde{l}(\gamma) \geqslant 0$ if and only if $l(a_m) \geqslant 0$, if and only if all $l(a_i) \geqslant 0$ by Lemma 36.6, if and only if γ is a p-adic algebraic integer.

We return now to the proof of Theorem 36.4.

Proof. We must show that $\tilde{l}$ has the three properties in (ii) of Section 36.1. Now for $\alpha, \beta \in E$

$$(1)\ \tilde{l}(\alpha\beta) = (1/n)l(N(\alpha\beta)) = (1/n)l(N(\alpha)N(\beta))$$
$$= (1/n)l(N(\alpha)) + (1/n)l(N(\beta)) = \tilde{l}(\alpha) + \tilde{l}(\beta).$$

Suppose that $\tilde{l}(\alpha) \leqslant \tilde{l}(\beta)$. Then $0 \leqslant \tilde{l}(\beta) - \tilde{l}(\alpha) = \tilde{l}(\beta/\alpha)$, by (1). But then β/α is a p-adic algebraic integer by Lemma 36.8, and hence $1 + \beta/\alpha$ is also a p-adic algebraic integer. Then, again by Lemma 36.8, $\tilde{l}(1 + \beta/\alpha) \geqslant 0$, so that

$$(2) \qquad \tilde{l}(\alpha + \beta) \geqslant \tilde{l}(\alpha) = \text{minimum}\{\tilde{l}(\alpha), \tilde{l}(\beta)\}.$$

Finally

$$(3) \qquad \tilde{l}(0) = (1/n)l(N(0)) = (1/n)l(0) = \infty.$$

Therefore $\tilde{l}$ has the desired properties (1) to (3). Moreover, if $a \in \Omega_p$, $e_i\rho(a) = ae_i$, $i = 1, 2, \ldots, n$.

$$\therefore \quad \rho(a) = \begin{pmatrix} a & & \\ & \ddots & \\ & & a \end{pmatrix} \quad \text{and} \quad N(a) = a^n.$$

$$\therefore \quad \tilde{l}(a) = (1/n)l(N(a)) = (1/n)l(a^n) = l(a),$$

showing that $\tilde{l}$ is an extension of l.

To conclude the proof we must show that E is complete with respect to $\tilde{l}$, or equivalently, with respect to the metric $\tilde{V} = c^{\tilde{l}}$, $0 < c < 1$. Now E is a vector space over Ω_p with the metric $\tilde{V}$ and we have the following result [2, p. 54]: *Any two metrics on a finite-dimensional vector space are equivalent*, *i.e.*, *a sequence which is a Cauchy sequence with respect to either metric is also a Cauchy sequence with respect to the other.*

Now, resuming the proof of Theorem 36.4, let

$$\alpha = a_1 e_1 + a_2 e_2 + \cdots + a_n e_n\,, \qquad a_i \in \Omega_p\,,$$

be an arbitrary element of E. Observe that $V^*(\alpha) = \max_i \{V(a_i)\}$ has the properties (1)–(3) of Section 36.2 and is therefore also a metric on E. Let

$$\beta_i = b_{i1} e_1 + b_{i2} e_2 + \cdots + b_{in} e_n \qquad (i = 1, 2, \ldots)$$

be a Cauchy sequence of E with respect to $\tilde{V}$. Then, for i, j sufficiently large

$$\tilde{V}(\beta_i - \beta_j) < \epsilon \Rightarrow V^*(\beta_i - \beta_j) < \epsilon \Rightarrow V(b_{ik} - b_{jk}) < \epsilon, \qquad k = 1, 2, \ldots, n.$$

Thus, for all k, the sequences $b_{1k}, b_{2k}, \ldots$ are Cauchy sequences and $\lim_{i\to\infty} b_{ik} = b_k \in \Omega_p$. If we now put $\beta = b_1 e_1 + b_2 e_2 + \cdots + b_n e_n$, then $\beta \in E$ and $\lim_{i\to\infty} V^*(\beta - \beta_i) = 0$, so that $\lim_{i\to\infty} \tilde{V}(\beta - \beta_i) = 0$ and this means that $\lim_{i\to\infty} \beta_i = \beta$. Thus E is complete.

37. Algebras over a p-Adic Field

1. Notation

- E A complete p-adic field; a finite extension of Ω_p which in turn is the p-adic completion of the rational field P.
- $\mathfrak{o}$ The principal ideal ring of p-adic integers of E.
- $\mathfrak{p}$ The maximal ideal in $\mathfrak{o}$.
- $\bar{\mathfrak{o}}$ $\mathfrak{o}/\mathfrak{p}$, the residue class ring; a field of characteristic p.
- p The unique rational prime in $\mathfrak{p}$.
- A An algebra, with identity, over E.
- $\mathfrak{O}$ An order in A (with respect to $\mathfrak{o}$).

By Theorem 34.6 any representation of A is similar to an

integral representation and we shall generally assume that any given representation μ is integral.

$A(x), B(x)$ Matrices depending on $x \in \mathfrak{O}$, with coefficients in $\mathfrak{o}$.

$\overline{A(x)}, \overline{B(x)}$ Corresponding matrices in $\bar{\mathfrak{o}}$, obtained by replacing each coefficient in $A(x)$, $B(x)$ by its residue class modulo $\mathfrak{p}$.

$\rho(x)$ The (integral) regular representation of $\mathfrak{O}$.

$\mu(x)$ An arbitrary (integral) representation of $\mathfrak{O}$.

$\bar{\sigma}(x), \bar{\mu}(x)$ Matrix representations of $\mathfrak{O}$ in $\bar{\mathfrak{o}}$.

2. Preliminary Results

These concern the implication, for an integral representation, of reducibility or decomposability in the corresponding modular representation under certain conditions.

(37.1) **Lemma.** *Let*

$$\mu(x) = \begin{pmatrix} A(x) & B(x) \\ C(x) & D(x) \end{pmatrix}, \qquad x \in \mathfrak{O},$$

be an integral representation of the order $\mathfrak{O}$ in A, and let

$$\overline{A(x)} = \bar{\mu}_1(x), \qquad \textit{a modular representation}$$

$$\overline{D(x)} = \bar{\rho}_+(x), \qquad \textit{a component of the modular regular representation}$$

$$\overline{B(x)} = 0$$

(in other words, the modular representation $\bar{\mu}(x) = \overline{\mu(x)}$ has a component of the modular regular representation as a bottom constituent). *Then there exists a matrix H in $\mathfrak{o}$ such that*

$$-HA(x) + C(x) + D(x)H \equiv 0 \qquad (\text{mod } \mathfrak{p}).$$

Proof. The modular representation $\bar{\mu}$ given by

$$\bar{\mu}(x) = \overline{\mu(x)} = \begin{pmatrix} \bar{\mu}_1(x) & 0 \\ \overline{C(x)} & \bar{\rho}_+(x) \end{pmatrix}$$

is reducible. By Corollary 6.6′ there is a matrix

$$\bar{T} = \begin{pmatrix} I & 0 \\ \bar{H} & I \end{pmatrix}$$

such that

$$(*) \qquad \bar{T}^{-1}\bar{\mu}(x)\bar{T} = \begin{pmatrix} \bar{\mu}_1(x) & 0 \\ 0 & \bar{\rho}_+(x) \end{pmatrix}.$$

Replacing each coefficient in the matrix H by a representative of its class in $\mathfrak{o}$ we get an integral matrix H and an integral matrix

$$T = \begin{pmatrix} I & 0 \\ H & I \end{pmatrix}.$$

Moreover,

$$T^{-1} = \begin{pmatrix} I & 0 \\ -H & I \end{pmatrix}$$

is also an integral matrix. Now

$$T^{-1}\mu(x)T =$$

$$\begin{pmatrix} A(x) + B(x)H & B(x) \\ -HA(x) + C(x) + HB(x)H + D(x)H & -HB(x) + D(x) \end{pmatrix}$$

so that going over to the modular statement and comparing with $(*)$ we get

$$-HA(x) + C(x) + D(x)H \equiv 0 \qquad (\mathrm{mod}\ \mathfrak{p}).$$

(37.2) **Lemma.** *Let $\mu(x)$ be an integral matrix representation of the order $\mathfrak{O}$ in A, such that the corresponding modular representation $\bar{\mu}(x) = \overline{\mu(x)}$ is the direct sum of a modular representation $\bar{\mu}_1(x)$ and a component $\bar{\rho}_+(x)$ of the modular regular representation. Thus $\mu(x)$ is of the form*

$$\mu(x) = \begin{pmatrix} A(x) & \mathfrak{p}B(x) \\ \mathfrak{p}C(x) & D(x) \end{pmatrix},$$

$$\overline{(Ax)} = \bar{\mu}_1(x), \qquad \overline{D(x)} = \bar{\rho}_+(x).$$

Then there exists an integral matrix

$$T = \begin{pmatrix} I & 0 \\ \mathfrak{p}L' & I \end{pmatrix}$$

such that

$$T^{-1}\mu(x)T = \begin{pmatrix} \mu_1(x) & \mathfrak{p}B(x) \\ 0 & \rho_+(x) \end{pmatrix}$$

in which $\mu_1(x)$, $\rho_+(x)$ *are integral representations of* $\mathfrak{O}$ *and*

$$\bar{\mu}_1(x) = \overline{\mu_1(x)} \qquad \bar{\rho}_+(x) = \overline{\rho_+(x)}.$$

Proof. The required matrix will be constructed as the limit T of a sequence $\{T_j\}$ of matrices. Let $T_0 = I$, the identity matrix, and assume that we already have a sequence of integral matrices:

$$(\alpha) \qquad T_1, \quad T_2, \ldots, T_{k-1}$$

such that

$$(1) \qquad T_j = \begin{pmatrix} I & 0 \\ L_j & I \end{pmatrix}, \qquad L_j \text{ an integral matrix,}$$

$$(2) \qquad T_j \equiv T_{j-1} \qquad (\text{mod } \mathfrak{p}^j), \qquad j = 1, 2, \ldots, k-1.$$

It follows from (2) that $T_j \equiv T_{j-1}\,(\text{mod } \mathfrak{p})$, $1 \leqslant j \leqslant k-1$, and so

$$(3) \qquad T_j \equiv I \qquad (\text{mod } \mathfrak{p}), \qquad 1 \leqslant j \leqslant k-1.$$

Set

$$(4) \qquad \mu_j(x) = T_j^{-1}\mu(x)T_j, \qquad j = 1, 2, \ldots, k-1.$$

Then $\mu_j(x)$ is an integral representation, and by (3)

$$(5) \qquad \mu_j(x) \equiv I^{-1}\mu(x)I = \mu(x) \qquad (\text{mod } \mathfrak{p}).$$

Recalling the form of $\mu(x)$ shown in the statement of the theorem we see that $\mu_j(x)$ has the form

$$(6) \qquad \mu_j(x) = \begin{pmatrix} A_j(x) & \mathfrak{p}B(x) \\ K_j(x) & D_j(x) \end{pmatrix}, \qquad j = 1, 2, \ldots, k-1,$$

where $A_j(x) = A(x) + \mathfrak{p}B(x)L_j$, $D_j(x) = D(x) - \mathfrak{p}L_jB(x)$, and we have made the further inductive assumption that

$$K_j(x) = \mathfrak{p}^{j+1}C_j(x), \qquad j = 1, 2, ..., k-1. \tag{7}$$

Note that $\overline{A_j(x)} = \overline{A(x)} = \bar{\mu}_1(x)$ and $\overline{D_j(x)} = \overline{D(x)} = \bar{\rho}_+(x)$. We now have the sequence (α) and the sequence

$$\mu_1, \quad \mu_2, ..., \mu_{k-1} \tag{β}$$

which are related by (4) and satisfy (1)–(7). It will now be shown that (α) and (β) can be extended by T_k and μ_k.

Since μ_j is a representation $\mu_j(x \pm y) = \mu_j(x) \pm \mu_j(y)$, $\mu_j(a) = aI$, $a \in \mathfrak{o}$ ($a \in \mathfrak{o}$ identified with $a \cdot 1 \in \mathfrak{O}$) and $\mu_j(xy) = \mu_j(x)\mu_j(y)$. It follows from these relations, using the form (6) for $\mu_j(x)$, that

$$\begin{aligned} C_j(x \pm y) &= C_j(x) \pm C_j(y), \\ C_j(a) &= 0, \qquad a \in \mathfrak{o}, \\ C_j(xy) &= C_j(x)A_j(y) + D_j(x)C_j(y). \end{aligned} \tag{8}$$

If we put

$$\hat{\mu}_j(x) = \begin{pmatrix} A_j(x) & 0 \\ C_j(x) & D_j(x) \end{pmatrix},$$

it follows from (8) that $\hat{\mu}_j(x)$ is an integral representation. Since

$$\overline{\hat{\mu}_j(x)} = \begin{pmatrix} \overline{A_j(x)} & 0 \\ \overline{C_j(x)} & \overline{D_j(x)} \end{pmatrix} = \begin{pmatrix} \bar{\mu}_1(x) & 0 \\ \overline{C_j(x)} & \bar{\rho}_+(x) \end{pmatrix},$$

Lemma 37.1 applies, and there is an integral matrix H_j such that

$$-H_jA_j(x) + C_j(x) + D_j(x)H_j \equiv 0 \quad (\mathrm{mod}\ \mathfrak{p}), \qquad j = 1, 2, ..., k-1. \tag{9}$$

Put

$$\begin{aligned} T_k &= T_{k-1} \begin{pmatrix} I & 0 \\ \mathfrak{p}^k H_{k-1} & I \end{pmatrix}, \\ \mu_k &= T_k^{-1}\mu(x)T_k. \end{aligned}$$

It is trivial that (1)–(5) still hold for T_k and μ_k. We must show that (6) along with (7) hold also. Now

$$\mu_k(x) = \begin{pmatrix} I & 0 \\ -\mathfrak{p}^k H_{k-1} & I \end{pmatrix} T_{k-1}^{-1}\mu(x)T_{k-1} \begin{pmatrix} I & 0 \\ \mathfrak{p}^k H_{k-1} & I \end{pmatrix}$$

and since $\mu_{k-1}(x) = T_{k-1}^{-1}\mu(x)T_{k-1}$ we have, using (6) and (7),

$$\mu_k(x) = \begin{pmatrix} A_k(x) & \mathfrak{p}B(x) \\ * & D_k(x) \end{pmatrix}$$

where

$$\begin{aligned} A_k(x) &= A_{k-1}(x) + \mathfrak{p}^{k+1}B(x)H_{k-1} \\ &= A(x) + \mathfrak{p}B(x)L_{k-1} + \mathfrak{p}^{k+1}B(x)H_{k-1} \\ &= A(x) + \mathfrak{p}B(x)(L_{k-1} + \mathfrak{p}^k H_{k-1}) \\ &= A(x) + \mathfrak{p}B(x)L_k . \end{aligned}$$

Similarly

$$D_k(x) = D(x) - \mathfrak{p}L_k B(x)$$

where we have defined

$$L_k = (L_{k-1} + \mathfrak{p}^k H_{k-1}).$$

Thus the form of A_j, D_j (line after (6)) is preserved. The $*$ stands for

$$-\mathfrak{p}^k(H_{k-1}A_{k-1}(x) + C_{k-1}(x) - \mathfrak{p}^{k+1}H_{k-1}B(x)H_{k-1} + D_{k-1}(x)H_{k-1}).$$

Now (9) shows that the term within the parentheses is divisible by $\mathfrak{p}$ and so $*$ can be written $\mathfrak{p}^{k+1}C_k(x)$. Then

$$\mu_k(x) = \begin{pmatrix} A_k(x) & \mathfrak{p}B(x) \\ \mathfrak{p}^{k+1}C_k(x) & D_k(x) \end{pmatrix}$$

and the sequences (α) and (β) have been extended.

We may now assume an infinite sequence T_1, T_2, ... satisfying (1)–(7). Because of (2) the sequence converges and then so does

$\{T_j^{-1}\mu(x)T_j\} = \{\mu_j(x)\}$. Let $\lim_{j\to\infty} T_j = T$, and $\lim_{j\to\infty} L_j = L$. Taking limits in (6) we get

$$(10) \quad T^{-1}\mu(x)T = \begin{pmatrix} A(x) + \mathfrak{p}B(x)L & \mathfrak{p}B(x) \\ 0 & D(x) - \mathfrak{p}LB(x) \end{pmatrix}.$$

Then the integral matrices $\mu_1(x) = A(x) + \mathfrak{p}B(x)L$, $\rho_+(x) = D(x) - \mathfrak{p}LB(x)$ are representations of $\mathfrak{O}$, and (10) has the form required by the Theorem with $\overline{\mu_1(x)} = \overline{A(x)} = \bar{\mu}_1(x)$ and similarly $\overline{\rho_+(x)} = \bar{\rho}_+(x)$. Finally, since $T_j \equiv I \pmod{p}$, for all j, and since

$$T_j = \begin{pmatrix} I & 0 \\ L_j & I \end{pmatrix}, \qquad L_j \equiv 0 \qquad (\text{mod } \mathfrak{p}).$$

Hence $L = \lim_{j\to\infty} L_j \equiv 0 \pmod{\mathfrak{p}}$. Thus $L = \mathfrak{p}L'$ and

$$T = \begin{pmatrix} I & 0 \\ \mathfrak{p}L' & I \end{pmatrix}.$$

The proof is now complete.

To summarize: This lemma shows that if a modular representation derived from an integral representation is decomposable and contains a component of the regular representation as a summand, then the integral representation itself is integrally equivalent to a reducible representation. If both summands of the modular representation are components of the regular representation, then a double application of the lemma proves that the integral representation is not only reducible but even decomposable. This is shown in Lemma 37.3.

Note that if ρ is the integral regular representation of an order $\mathfrak{O}$ in A, then the modular representation $\bar{\rho}(x) = \overline{\rho(x)}$ is called the modular regular representation. It is the regular representation of the $\bar{\mathfrak{o}}$ algebra $\bar{\mathfrak{O}}$ (Section 35, end of part 1).

Let $\nu(x)$ be an integral representation of $\mathfrak{O}$, and let $\overline{\nu(x)}$ be the corresponding modular representation. Suppose that

$$\overline{\nu(x)} \sim \begin{pmatrix} \bar{\nu}_1(x) & 0 \\ 0 & \bar{\nu}_2(x) \end{pmatrix}.$$

Then there is a matrix $\bar{P}$ in $\bar{\mathfrak{o}}$ such that

$$\bar{P}^{-1}\overline{\nu(x)}\bar{P} = \begin{pmatrix} \bar{\nu}_1(x) & 0 \\ 0 & \bar{\nu}_2(x) \end{pmatrix}.$$

Replacing each coefficient in $\bar{P}$ by a representative of its class in $\mathfrak{o}$, we get an integral matrix P. Since $\det \bar{P} \neq 0$, then $\det P \not\equiv 0 \pmod{\mathfrak{p}}$. Hence $\det P$ is a unit of $\mathfrak{o}$, so that $(\det P)^{-1} \in \mathfrak{o}$. Thus P^{-1} is an integral matrix. If we put $\sigma(x) = P^{-1}\nu(x)P$, then $\sigma(x)$ is an integral matrix similar to $\nu(x)$ and

$$\overline{\sigma(x)} = \bar{P}^{-1}\overline{\nu(x)}\,\bar{P} = \begin{pmatrix} \bar{\nu}_1(x) & 0 \\ 0 & \bar{\nu}_2(x) \end{pmatrix}.$$

Thus, if an integral representation ν has its corresponding modular representation similar to a decomposable representation, then ν is integrally equivalent to an integral representation σ whose corresponding modular representation is immediately decomposable.

(37.3) **Lemma.** *Let $\nu(x)$ be an integral representation of an order $\mathfrak{D}$ in A. If the modular representation $\bar{\nu}(x) = \overline{\nu(x)}$ is decomposable, $\bar{\nu}(x) \sim \bar{\rho}_1(x) \oplus \bar{\rho}_2(x)$, and $\bar{\rho}_1(x)$, $\bar{\rho}_2(x)$ are components of the regular representation, then $\nu(x)$ is decomposable:*

$$\nu(x) \sim \rho_1(x) \oplus \rho_2(x) \qquad \textit{and} \qquad \overline{\rho_1(x)} = \bar{\rho}_1(x), \qquad \overline{\rho_2(x)} = \bar{\rho}_2(x).$$

Proof. By the remarks preceding the statement of the lemma $\nu(x)$ may be taken at the outset so that $\bar{\nu}(x) = \overline{\nu(x)} = \bar{\rho}_1(x) \oplus \bar{\rho}_2(x)$. Then $\nu(x)$ has the form

$$\nu(x) = \begin{pmatrix} A(x) & \mathfrak{p}B(x) \\ \mathfrak{p}C(x) & D(x) \end{pmatrix}, \qquad \overline{A(x)} = \bar{\rho}_1(x), \qquad \overline{D(x)} = \bar{\rho}_2(x).$$

By Lemma 37.2 there is a matrix

$$T = \begin{pmatrix} I & 0 \\ \mathfrak{p}L & I \end{pmatrix}$$

such that

$$T^{-1}\nu(x)T = \begin{pmatrix} \nu_1(x) & \mathfrak{p}B(x) \\ 0 & \rho_2(x) \end{pmatrix}, \qquad \overline{\rho_2(x)} = \bar{\rho}_2(x), \qquad \overline{\nu_1(x)} = \bar{\rho}_1(x).$$

But then

$$\begin{pmatrix} 0 & I \\ I & 0 \end{pmatrix} T^{-1}\nu(x)T \begin{pmatrix} 0 & I \\ I & 0 \end{pmatrix} = \begin{pmatrix} \rho_2(x) & 0 \\ pB(x) & \nu_1(x) \end{pmatrix}.$$

Since $\overline{\nu_1(x)} = \bar{\rho}_1(x)$, the lemma can be used again to provide a matrix S such that

$$S^{-1} \begin{pmatrix} 0 & I \\ I & 0 \end{pmatrix} T^{-1}\nu(x)T \begin{pmatrix} 0 & I \\ I & 0 \end{pmatrix} S = \begin{pmatrix} \rho_2(x) & 0 \\ 0 & \rho_1(x) \end{pmatrix}$$

where we have written $\rho_1(x)$ for $\nu_1(x)$. Putting

$$P = T \begin{pmatrix} 0 & I \\ I & 0 \end{pmatrix} S \begin{pmatrix} 0 & I \\ I & 0 \end{pmatrix}$$

we get

$$P^{-1}\nu(x)P = \begin{pmatrix} \rho_1(x) & 0 \\ 0 & \rho_2(x) \end{pmatrix}, \qquad \overline{\rho_1(x)} = \bar{\rho}_1(x), \qquad \overline{\rho_2(x)} = \bar{\rho}_2(x),$$

which proves the lemma.

If $\bar{\rho}_1(x)$ and $\bar{\rho}_2(x)$ are themselves decomposable, $\bar{\rho}_1(x) \sim \bar{\rho}_{11}(x) \oplus \bar{\rho}_{12}(x)$ and $\bar{\rho}_2(x) \sim \bar{\rho}_{21}(x) \oplus \bar{\rho}_{22}(x)$, then since the $\bar{\rho}_{ij}(x)$ are necessarily components of the modular regular representation, the lemma applies again and we have in the same way

$$\nu(x) \sim \begin{pmatrix} \rho_{11}(x) & & & \\ & \rho_{12}(x) & & \\ & & \rho_{21}(x) & \\ & & & \rho_{22}(x) \end{pmatrix}, \qquad \overline{\rho_{ij}(x)} = \bar{\rho}_{ij}(x).$$

Continuing in this way we have in the special case in which $\nu(x) = \rho(x)$ is the integral regular representation:

(37.4) **Theorem.** *The regular representation ρ of an order $\mathfrak{O}$ in A, an algebra over a complete p-adic field E, is integrally equivalent to a representation $\rho'(x) = \sum_{i=1}^{l} \rho_i(x)$, such that $\overline{\rho'(x)} = \sum_{i=1}^{l} \overline{\rho_i(x)}$ is any prescribed decomposition of the modular regular representation into indecomposable components.*

38. A Connection between the Intertwining Numbers

Let $\mu(x)$ and $\nu(x)$ be two integral matrix representations of an order $\mathfrak{O}$ in A; let $\bar{\mu}(x) = \overline{\mu(x)}$ and $\bar{\nu}(x) = \overline{\nu(x)}$ be the corresponding modular representations. How are the intertwining numbers $i(\mu, \nu)$ and $i(\bar{\mu}, \bar{\nu})$ related? In a special case the answer is provided by:

(38.1) **Lemma.** *If* $\overline{\mu(x)}$ *is an indecomposable component of the modular regular representation, that is, a principal indecomposable component, then* $i(\mu, \nu) = i(\bar{\mu}, \bar{\nu})$.

Proof. (a) Let $\bar{T}$ be a matrix such that $\overline{\mu(x)}\bar{T} = \bar{T}\overline{\nu(x)}$. Replacing each coefficient in $\bar{T}$ by a representative of its class in $\mathfrak{o}$ we get a matrix T and

$$\mu(x)T \equiv T\nu(x) \qquad (\text{mod } \mathfrak{p}). \tag{1}$$

Let

$$\tau(x) = \begin{pmatrix} I & 0 \\ -T & 0 \end{pmatrix}\begin{pmatrix} \nu(x) & 0 \\ 0 & \mu(x) \end{pmatrix}\begin{pmatrix} I & 0 \\ T & I \end{pmatrix} = \begin{pmatrix} \nu(x) & 0 \\ \mu(x)T - T\nu(x) & \mu(x) \end{pmatrix}.$$

Because of (1) this can be written

$$\tau(x) = \begin{pmatrix} \nu(x) & 0 \\ \mathfrak{p}C(x) & \mu(x) \end{pmatrix}, \qquad \overline{\mu(x)} = \bar{\mu}(x), \quad \text{a component of } \bar{\rho}.$$

Then Lemma 37.2 applies and there is a matrix

$$H = \begin{pmatrix} I & 0 \\ \mathfrak{p}L & I \end{pmatrix}$$

such that

$$H^{-1}\tau(x)H = \begin{pmatrix} \nu(x) & 0 \\ 0 & \mu(x) \end{pmatrix}$$

$$\begin{pmatrix} \nu(x) & 0 \\ -\mathfrak{p}L\nu(x) + \mathfrak{p}C(x) + \mathfrak{p}\mu(x)L = & \mu(x) \end{pmatrix}.$$

Hence

$$\text{(2)} \qquad -\mathfrak{p}L\nu(x) + \mathfrak{p}C(x) + \mathfrak{p}\mu(x)L = 0.$$

But $\mathfrak{p}C(x)$ was written for $\mu(x)T - T\nu(x)$. Thus from (2)

$$\mu(x)(T + \mathfrak{p}L) = (T + \mathfrak{p}L)\nu(x).$$

Put $P = T + \mathfrak{p}L$. Since $\bar{P} = \bar{T}$, we see that every matrix $\bar{T}$ intertwining $\bar{\mu}$ and $\bar{\nu}$ derives from an integral matrix P intertwining μ and ν.

(b) Let $\mathfrak{M} = \{T : \mu(x)T = T\nu(x)\}$. Then $\mathfrak{M}$ is an E-module of matrices over E. Recall that if $\alpha \in E$, then $\alpha = a/b$, $a \in \mathfrak{o}$, $b \in \mathfrak{o}$. Also $i(\mu, \nu) =$ rank of $\mathfrak{M}$, by definition. Suppose that $i(\mu, \nu) = r$. Let $T_1, T_2, \ldots, T_r$ be an E-basis of $\mathfrak{M}$. For every j, let l_j be the least common multiple of the denominators of all the coefficients in the matrix T_j. Then $T_j' = l_jT_j$ is an integral matrix and the set $T_1', \ldots, T_r'$ is linearly independent over E.

Let $\mathfrak{N}$ be the set of all integral matrices in $\mathfrak{M}$. Then $\mathfrak{N}$ is a finitely generated $\mathfrak{o}$-module and contains the T_j'. Hence by Lemma 34.1 there are integral matrices $P_1, \ldots, P_r$ which $\mathfrak{o}$-generate $\mathfrak{N}$ and are linearly independent over E. If $\bar{T}$ is any matrix intertwining $\bar{\mu}$ and $\bar{\nu}$, we know from part (a) that there is an integral matrix P such that $\bar{P} = \bar{T}$. Since the P_i are an $\mathfrak{o}$-basis of $\mathfrak{N}$,

$$P = a_1P_1 + \cdots + a_rP_r, \qquad a_i \in \mathfrak{o},$$

and

$$\bar{T} = \bar{a}_1\bar{P}_1 + \cdots + \bar{a}_r\bar{P}_r, \qquad \bar{a}_i \in \bar{\mathfrak{o}}.$$

Thus $\bar{P}_1, \ldots, \bar{P}_r$ are generators of the module $\bar{\mathfrak{M}}$ of all matrices intertwining $\bar{\mu}$ and $\bar{\nu}$. Hence $i(\bar{\mu}, \bar{\nu}) \leqslant r$.

To complete the proof we must show that the $\bar{P}_i$ are linearly independent. Suppose that $\bar{c}_1\bar{P}_1 + \cdots + \bar{c}_r\bar{P}_r = 0$. Then there are elements $c_i \in \mathfrak{o}$ and an integral matrix H such that

$$c_1P_1 + \cdots + c_rP_r = \mathfrak{p}H.$$

$$\therefore \quad \frac{c_1}{\mathfrak{p}}P_1 + \cdots + \frac{c_r}{\mathfrak{p}}P_r = H.$$

Since H is an integral matrix and the P_i are an $\mathfrak{o}$-basis of $\mathfrak{N}$, $c_j/\mathfrak{p} \in \mathfrak{o}$, $j = 1, 2, \ldots, r$, so that $\mathfrak{p} \mid c_j$ or $\bar{c}_j = 0$. Thus $\{\bar{P}_j\}$ are a basis of $\overline{\mathfrak{M}}$ and $i(\bar{\mu}, \bar{\nu}) = r = i(\mu, \nu)$.

39. Modular Representations of Groups

Let $\rho_1, \rho_2, \ldots, \rho_s$ be the absolutely irreducible representations of a finite group G in an algebraically closed field F of characteristic zero. Then F contains P, the field of rational numbers. Since the group is of finite order n the set of matrices $\{\rho_i(g)\}$, for all $g \in G$ and all irreducible representations ρ_i, is finite. Adjoining to P the finite set of coefficients of all these matrices we get a field B in which all the representations lie. Let us ensure, by a further adjunction if necessary, that B contains the nth roots of unity. From the theory of field extensions we know that B can be obtained by the adjunction of a single element $\theta \in F$, which satisfies a unique irreducible equation

$$(39.1) \quad x^m + a_1 x^{m-1} + \cdots + a_{m-1} x + a_m = 0, \qquad a_i \in P,$$

where m is the degree of $B = P(\theta)$ over P.

Similarly, if Σ is an algebraically closed field of characteristic p and Π is the prime field of p elements we have that the absolutely irreducible representations $\bar{\rho}_1, \bar{\rho}_2, \ldots, \bar{\rho}_l$ of G in Σ lie in a field $\Gamma = \Pi(\bar{\eta})$, where $\bar{\eta}$ satisfies a unique irreducible equation

$$(39.2) \qquad x^d + \bar{c}_1 x^{d-1} + \cdots + \bar{c}_d = 0, \qquad \bar{c}_i \in \Pi.$$

Then Γ is a finite field of p^d elements.

Now let J_B be the ring of integers of B. These are all elements of B which satisfy an equation $x^t + b_1 x^{t-1} + \cdots + b_t = 0$, in which the b_i are rational integers.

We know from the theory of integral algebraic domains [16, pp. 134–144; 22, pp. 75–89] that:

(1) J_B is a ring.

(2) B is the quotient field of J_B.

(3) J_B has a basis $\theta_1, \theta_2, ..., \theta_m$ such that if $\alpha \in J_B$ then

$$\alpha = k_1\theta_1 + k_2\theta_2 + \cdots + k_m\theta_m$$

and the k_i are unique rational integers.

(4) Every ideal $\mathfrak{A}$ of J_B has a linearly independent (over B) basis $\omega_1, ..., \omega_m$ such that if $\alpha \in \mathfrak{A}$, then

$$\alpha = k_1\omega_1 + k_2\omega_2 + \cdots + k_m\omega_m$$

and again the k_i are unique rational integers.

In (3) and (4) the integer m is the degree of $P(\theta)$ over P.

Let $\{p\}$ be a maximal ideal of J_B containing the prime p. Then $\Delta \cong J_B/\{p\}$ is a field of characteristic p. If we denote the residue class of α by $[\alpha]$ we have, from 3,

$$[\alpha] = [k_1][\theta_1] + \cdots + [k_m][\theta_m], \qquad [k_i] \in \Pi,$$

so that Δ has at most p^m elements.

We ensure that Δ contains Γ by having B contain from the beginning a root η of an equation

(39.3)
$$x^d + c_1x^{d-1} + \cdots + c_d = 0, \qquad c_i \text{ integers}, \qquad 0 \leqslant c_i \leqslant \mathfrak{p} - 1,$$

which, taken modulo $\{p\}$, yields (39.2).

Finally, we embed B in a complete p-adic field E by considering (39.1) as an equation over the p-adic field $\Omega_p \supset P$. In this way B lies in the algebraic extension E of Ω_p. Moreover, because of Lemma 36.8 the ring $\mathfrak{o}$ of p-adic integers of E contains J_B.

(39.1) ***Lemma.*** $\mathfrak{o}/\mathfrak{p} = \bar{\mathfrak{o}} \cong \Delta = J_B/\{p\}$.

Proof. Recalling that $\mathfrak{p}$ is the maximal prime ideal in $\mathfrak{o}$ and contains p and that $J_B \subset \mathfrak{o}$, we see that $\{p\} = J_B \cap \mathfrak{p}$. Also the ideal sum $(J_B, \mathfrak{p})$ containing $\mathfrak{p}$ and J_B is $\mathfrak{o}$, by the maximality of $\mathfrak{p}$.

Then by the third isomorphism theorem applied to ideals and envisioned in the diagram

$$\begin{array}{ccc} & \mathfrak{o} = (J_B, \mathfrak{p}) & \\ / & & \diagdown \\ J_B & & \mathfrak{p} \\ \diagdown & & / \\ & \{p\} = J_B \cap \mathfrak{p} & \end{array}$$

we have $J_B/\{p\} \cong \mathfrak{o}/\mathfrak{p}$.

This lemma shows that the modular representation in Δ derived from (integral) representations in J_B can be studied as representations in $\bar{\mathfrak{o}}$ derived from representations in $\mathfrak{o}$. Since $\mathfrak{o}$ is a principal ideal ring the results developed for such a case can be applied.

40. Cartan Invariants and Decomposition Numbers

Notation

$$E, \quad \mathfrak{o}, \quad \mathfrak{p}, \quad p, \quad B, \quad J_B, \quad P, \quad \bar{\mathfrak{o}},$$

and the use of the overline are as in the preceding sections [Section 37].

Note that E, a complete p-adic field, is a finite extension of the field Ω_p and contains the field B, which in turn is a finite extension of the rational field P sufficiently large to contain all the absolutely irreducible representations of the finite group G.

ρ The regular representation of G in B.

$\rho_1, \rho_2, \ldots, \rho_s$ Principal indecomposable representations of G in B.

$\sigma_1, \sigma_2, \ldots, \sigma_s$ Irreducible representations of G.

We know (Theorem 7.2) that the number s is the same in both cases and that σ_j is the unique irreducible bottom constituent of ρ_j.

∂_j Any component of ρ such that $\bar{\partial}_j$ is a principal indecomposable representation of G in $\bar{\mathfrak{o}}$.

By Lemma 37.3 and Theorem 37.4 every $\bar{\partial}_j$ is obtainable from a ∂_j.

$\bar{\partial}_1, \bar{\partial}_2, ..., \bar{\partial}_l$ Principal indecomposable representations of G in $\bar{\mathfrak{o}}$.

$\bar{\tau}_1, \bar{\tau}_2, ..., \bar{\tau}_l$ Irreducible representations of G in $\bar{\mathfrak{o}}$. (Theorem 7.2 applies here too.)

Recall [Section 39] that $\bar{\mathfrak{o}}$ ($\simeq \Delta$) is large enough to accommodate all the absolute irreducible modular representations.

(40.1) **Definition.** *The Cartan invariant c_{ij} is the nonnegative integer expressing the multiplicity with which the irreducible modular representation $\bar{\tau}_j$ occurs as a constituent of the principal indecomposable modular representation $\bar{\partial}_i$:*

(40.2)

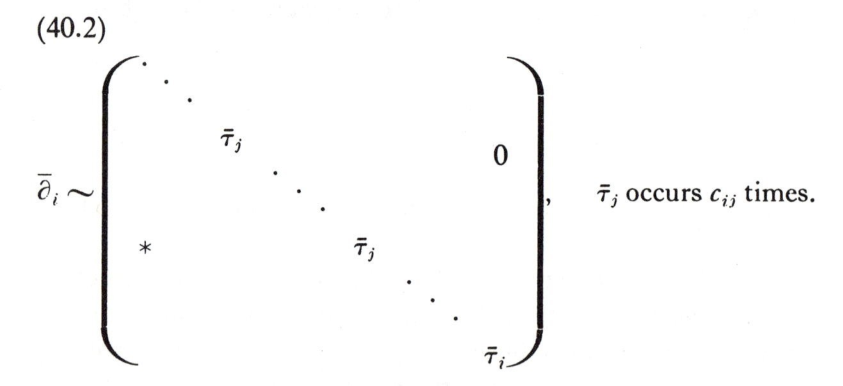

By Theorem 11.10 (with $k = 1$ by Lemma 11.9)

(40.3) $$c_{ij} = i(\bar{\partial}_j, \bar{\partial}_i),$$

the latter being the intertwining number, or the rank of the module of matrices intertwining $\bar{\partial}_j$ with $\bar{\partial}_i$.

Let $C = (c_{ij})$. Then C is an $l \times l$ matrix of integers.

(40.4) **Definition.** *The decomposition number d_{ij} is the multiplicity of the modular irreducible representation $\bar{\tau}_j$ in $\bar{\sigma}_i$, where $\bar{\sigma}_i$ is the* *modular representation determined by the ordinary (integral) irreducible representation* σ_i:

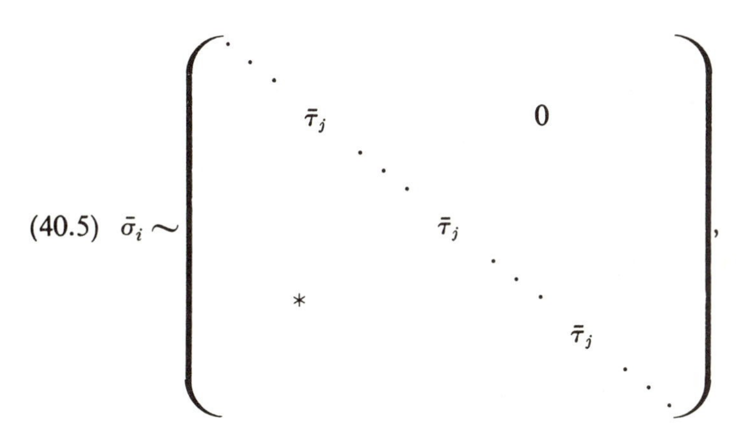

$$\bar{\sigma}_i \sim \begin{pmatrix} \ddots & & & & & & 0 \\ & \bar{\tau}_j & & & & & \\ & & \ddots & & & & \\ & & & \bar{\tau}_j & & & \\ & & & & \ddots & & \\ & * & & & & \bar{\tau}_j & \\ & & & & & & \ddots \end{pmatrix}, \tag{40.5}$$

$\bar{\tau}_j$ occuring d_{ij} times.

Note that though σ_i is irreducible, it cannot be assumed that $\bar{\sigma}_i$ is also irreducible. Writing $D = (d_{ij})$ we see that D is an $s \times l$ matrix of integers. Again, by Theorem 11.10,

$$\begin{aligned} d_{ij} &= i(\bar{\partial}_j , \bar{\sigma}_i) \\ &= i(\partial_j , \sigma_i) \qquad \text{by Lemma 38.1.} \end{aligned} \tag{40.6}$$

Now ∂_j as a representation in the field E of characteristic zero is completely reducible: $\partial_j \sim \sum_{k=1}^{s} h_{kj}\sigma_k$, h_{kj} , integers $\geqslant 0$. Then

$$d_{ij} = i(\partial_j , \sigma_i) = i\left(\sum_{k=1}^{s} h_{kj}\sigma_k , \sigma_i\right) = \sum_{k=1}^{s} h_{kj}i(\sigma_k , \sigma_i) = h_{ij} ,$$

the last step because only the zero matrix intertwines distinct irreducible representations and only the identity matrix commutes with an irreducible representation. Thus

$$\partial_j \sim \sum_{i=1}^{s} d_{ij}\sigma_i . \tag{40.7}$$

From (40.6) and (40.7) we have:

(40.9) **Lemma.** *$\bar{\sigma}_i$ contains $\bar{\tau}_j$ with the same multiplicity, d_{ij}, that ∂_j contains σ_i.*

Finally

$$c_{ij} = i(\bar{\partial}_j, \bar{\partial}_i) = i(\partial_j, \partial_i), \qquad \text{by Lemma 38.1}$$

$$= i\Big(\sum_{\lambda=1}^{s} d_{\lambda j}\sigma_\lambda, \sum_{\mu=1}^{s} d_{\mu i}\sigma_\mu\Big), \qquad \text{by (40.7)}$$

$$= \sum_{\lambda,\mu=1}^{s} d_{\lambda j} d_{\mu i} i(\sigma_\lambda, \sigma_\mu)$$

$$= \sum_{\lambda=1}^{s} d_{\lambda j} d_{\lambda i} = \sum_{\lambda=1}^{s} d_{\lambda i} d_{\lambda j}.$$

Hence

(40.10) $$c_{ij} = \sum_{\lambda=1}^{s} d_{\lambda i} d_{\lambda j}$$

and

(40.11) $$c_{ij} = c_{ji}.$$

Thus the matrix C is symmetric, and (40.10) shows that

$$C = D'D$$

where D' is the transpose of D.

Remarks. It can be shown [10, p. 602] that the determinant of the Cartan matrix C is a power of p and that the elementary divisors of D, the decomposition matrix, are all 1.

41. Character Relations

Let

$\chi^{\sigma_i} = \chi^i$ denote the character of the ordinary irreducible representation σ_i, $i = 1, 2, \ldots, s$;

ϕ_j denote the Brauer character of the irreducible modular representation $\bar{\tau}_j$, $j = 1, 2, \ldots, l$;

ψ_j denote the Brauer character of the principal indecomposable modular representation $\bar{\partial}_j$, $j = 1, 2, \ldots, l$.

It will be recalled that these are all complex-valued functions and that the last two are defined only on the p-regular elements of the group G. Moreover, from the definition of the Brauer character it follows that $\chi^{\sigma} = \chi^{\bar{\sigma}}$ and $\chi^{\partial} = \chi^{\bar{\partial}}$ for p-regular elements.

From (40.7) we have $\chi^{\partial_j} = \sum_{i=1}^{s} d_{ij}\chi^{\sigma_i}$, and so

$$\psi_j = \sum_{i=1}^{s} d_{ij}\chi^i, \qquad 1 \leqslant j \leqslant l. \tag{41.1}$$

Similarly from (40.2) and (40.5) we have

$$\psi_i = \sum_{j=1}^{l} c_{ij}\phi_j\,, \qquad 1 \leqslant i \leqslant l, \tag{41.2}$$

$$\chi^i = \sum_{j=1}^{l} d_{ij}\phi_j\,, \qquad 1 \leqslant i \leqslant s. \tag{41.3}$$

Let $C_1, C_2, \ldots, C_s$ be the classes of conjugate elements of G and let $\check{g}_j$ denote a fixed representative of the class C_j. Arrange the classes so that the first $m < s$ are p-regular. Let n be the order of the group and let h_j be the number of elements in the class C_j.

Form the matrices

$$\chi = \begin{pmatrix} \chi^1(\check{g}_1) & \chi^1(\check{g}_2) & \cdots & \chi^1(\check{g}_m) \\ \vdots & \vdots & \vdots\vdots\vdots & \vdots \\ \chi^s(\check{g}_1) & \chi^s(\check{g}^2) & \cdots & \chi^s(\check{g}_m) \end{pmatrix}, \qquad \text{an } s \times m \text{ matrix},$$

$$\phi = \begin{pmatrix} \phi_1(\check{g}_1) & \phi_1(\check{g}_2) & \cdots & \phi_1(\check{g}_m) \\ \vdots & \vdots & \vdots\vdots\vdots & \vdots \\ \phi_l(\check{g}_1) & \phi_l(\check{g}_2) & \cdots & \phi_l(\check{g}_m) \end{pmatrix}, \qquad \text{an } l \times m \text{ matrix},$$

$$\psi = \begin{pmatrix} \psi_1(\check{g}_1) & \psi_1(\check{g}_2) & \cdots & \psi_1(\check{g}_m) \\ \vdots & \vdots & \vdots\vdots\vdots & \vdots \\ \psi_l(\check{g}_1) & \psi_l(\check{g}_2) & \cdots & \psi_l(\check{g}_m) \end{pmatrix}, \qquad \text{an } l \times m \text{ matrix}.$$

Then since $D = (d_{ij})$ and $C = (c_{ij})$ the relations (41.1)–(41.3) yield

(41.1′) $$D'\chi = \psi, \qquad D' = \text{transpose of } D,$$

(41.2′) $$C\phi = \psi,$$

(41.3′) $$D\phi = \chi.$$

Let us introduce the $m \times m$ matrix $T = (t_{ij})$ where $t_{ij} = 1$ or 0 according as $\breve{g}_i$ and $\breve{g}_j^{-1}$ are or are not p-regular conjugate elements. Clearly $\delta_{ij} = t_{i\lambda}t_{\lambda j}$, since each summand is zero unless $\breve{g}_i \sim \breve{g}_\lambda^{-1}$ and $\breve{g}_\lambda \sim \breve{g}_j^{-1}$, that is, unless $\breve{g}_i \sim \breve{g}_j$ and so $i = j$, in which case a single summand $=1$. Hence $T^2 = I$, the identity matrix, and T is nonsingular.

Now the orthogonality relation (19.9) for the χ^i,

$$\sum_{\lambda=1}^{s} \chi^\lambda(\breve{g}_i)\chi^\lambda(\breve{g}_j^{-1}) = \delta_{ij}n/h_i ,$$

can be written

$$\sum_{\kappa=1}^{m}\sum_{\lambda=1}^{s} \chi^\lambda(\breve{g}_i)\chi^\lambda(\breve{g}_\kappa)t_{\kappa j} = \delta_{ij}n/h_i$$

or, in matrix form: $\chi'\chi T = (\delta_{ij}n/h_i)$. Since $T^2 = I$ this is

(41.4) $$\chi'\chi = (t_{ij}n/h_i) = \tilde{T}.$$

Taking the transpose in (41.3′) and multiplying on the right by χ we get, after using (41.4),

(41.5) $$\phi'\psi = (t_{ij}n/h_i) = \tilde{T}.$$

Now the $m \times m$ matrix on the right is nonsingular, since its square is the diagonal matrix $(n^2/h_1^2 , \ldots, n^2/h_m^2)$. Therefore

$$m = \operatorname{rank} \phi'\psi \leqslant \operatorname{rank} \phi' \leqslant m,$$

and thus

$$m = \operatorname{rank} \phi' = \operatorname{rank} \phi.$$

Since the matrix ϕ has l rows we have $l \geqslant m$.

Now l is the number of irreducible modular representations and m is the number of p-regular classes in G. We prove:

(41.6) **Lemma.** *The number of absolutely irreducible representations of a finite group G in a modular field of characteristic p is equal to the number of p-regular classes of conjugate elements of G.**

Proof. We have seen that $l \geqslant m$. Assume that $l > m$. Then there is a nontrivial relation among the rows of ϕ:

$$(*) \qquad \sum_{i=1}^{l} y_i \phi_i(\check{g}_j) = 0, \qquad j = 1, 2, ..., m,$$

and not all $y_i = 0$. Recall that all elements here lie in a field B which is a sufficiently large algebraic extension of the rational field P, and which is embedded in a complete p-adic field E. Moreover, the $\phi_i(\check{g}_j)$, as sums of roots of unity, belong to $\mathfrak{o}$, the ring of integers of E. Now E is the quotient field of $\mathfrak{o}$ so that we can ensure that the y_i in $(*)$ are integers by multiplying through by a suitable integer. Finally, we may assume that not all y_i are divisible by $\mathfrak{p}$, since any common factor $\mathfrak{p}$ may be canceled out.

Taking $(*)$ modulo $\mathfrak{p}$ we write $\sum_{i=1}^{l} \bar{y}_i \overline{\phi_i(\check{g}_j)} = 0$, in the field $\bar{\mathfrak{o}}$. But $\overline{\phi_i(\check{g}_j)} = \text{trace}\, \bar{\tau}_i(\check{g}_j)$, since in the first place the Brauer character ϕ_i was obtained from the trace of $\bar{\tau}_i$ by replacing roots of unity in $\bar{\mathfrak{o}}$ by the corresponding roots of unity in $\mathfrak{o}$. Thus

$$\sum_{i=1}^{l} \bar{y}_i \,\text{trace}\, \bar{\tau}_i(\check{g}_j) = 0, \qquad j = 1, 2, ..., m.$$

Since the matrix representation of an element g_j and of its p-regular factor f_j have the same characteristic roots (Lemma 31.3), trace $\bar{\tau}_i(g_j) =$ trace $\bar{\tau}_i(f_j)$ and the equation above can be extended so that

$$\sum_{i=1}^{l} \bar{y}_i \,\text{trace}\, \bar{\tau}_i(g) = 0, \qquad \forall g \in G.$$

* This theorem can be proved more directly, without the present apparatus of integral representations [4, 24, 18].

Then by the linearity of $\bar{\tau}_i$ and of the trace function the relation holds for any element $a = \sum_{i=1}^{n} \alpha_i g_i$ of the group algebra $A(G, \bar{\mathfrak{o}})$:

$$\sum_{i=1}^{l} \bar{y}_i \operatorname{trace} \bar{\tau}_i(a) = 0, \qquad \forall a \in A(G, \bar{\mathfrak{o}}).$$

Now for every i, the theorem of Frobenius and Schur shows that there is an element a such that

$$\bar{\tau}_i(a) = \begin{pmatrix} 1 & 0 & \cdot \\ 0 & 0 & \\ \cdot & & 0 \end{pmatrix}, \qquad \bar{\tau}_j(a) = 0,$$

giving, by the last equation, $\bar{y}_i = 0$. This is a contradiction so that $l = m$ and the theorem is proved.

Since we had $m = \operatorname{rank} \phi = \operatorname{rank} \psi$, we have:

(41.7) **Corollary.** *ϕ and ψ are nonsingular $l \times l$ matrices.*

Remarks. The $s \times l$ matrix χ has rank l as can be seen from (41.4), remembering that $m = l$. Taking the coefficients of χ modulo $\mathfrak{p}$ we get an $s \times l$ matrix $\bar{\chi}$ in $\bar{\mathfrak{o}}$. An important result [4] states that $\bar{\chi}$ has rank l. By (41.3″) $\chi = D\phi$ and hence $\bar{\chi} = \bar{D}\bar{\phi}$. Then $l = \operatorname{rank} \bar{\chi} = \operatorname{rank} \bar{D}\bar{\phi} \leqslant \operatorname{rank} \bar{\phi} \leqslant l$, so that $l = \operatorname{rank} \bar{\phi}$ and determinant $\bar{\phi} \neq 0$. Thus $\det \phi \not\equiv 0 \pmod{\mathfrak{p}}$.

Equating the i, jth entry on both sides of (41.5) we get

$$\text{(41.7)} \qquad \sum_{\lambda=1}^{l} \phi_\lambda(\breve{g}_i)\psi_\lambda(\breve{g}_j) = t_{ij} n / h_i \,.$$

Now for fixed j and $i = 1, 2, \ldots, l$ this is a system of l equations for $\psi_1(\breve{g}_j), \ldots, \psi_l(\breve{g}_j)$.

Writing $\phi_{\lambda i} = \phi_\lambda(\breve{g}_i)$ and $\phi^{\lambda i}$ for the cofactor of $\phi_{\lambda i}$, Cramer's rule gives

$$(*) \qquad \det \phi \, \psi_\lambda(\breve{g}_j) = \sum_{i=1}^{l} t_{ij}(n/h_i)\phi^{\lambda i}.$$

Since $t_{ij} = 0$ unless $\check{g}_i \sim \check{g}_j^{-1}$ in which event $t_{ij} = 1$ and $h_i = h_j$, this yields

$$\det \phi\psi_\lambda(\check{g}_j) = (n/h_j)b, \qquad b \text{ some integer in } \mathfrak{o}.$$

Because of $(*)$ the highest power of p which divides n/h_j also divides $\psi_\lambda(\check{g}_j)$. In particular, if the group order $n = p^a c$, $(c, p) = 1$, then p^a is a divisor of $\psi_\lambda(1) =$ the degree of ψ_λ.

We have denoted the exponential valuation on E by $l(x)$. Then if $l(x)$ is normalized so that $l(p) = 1$ we have

$$l(\psi_\lambda(\check{g}_j)) \geqslant l(n/h_j), \qquad l(dg\psi_\lambda) \geqslant l(n) = a.$$

Finally, we mention the following:

Theorem. (Brauer [5, 6]). *The elementary divisors of the Cartan matrix C are the numbers $p^{l(n/h_i)}$, $i = 1, 2, \ldots, l$.*

42. Modular Orthogonality Relations

By manipulating the relations (41.2′) and (41.5) certain orthogonal relations between the modular characters can be stated explicitly.

Let us recall that $\tilde{T} = (t_{ij}n/h_i)$ where t_{ij} equals 1 or 0 according as the p-regular elements $\check{g}_i$ and $\check{g}_j^{-1}$ belong to the same or different conjugate classes, n is the order of the group, and h_i the number of elements in the conjugate class C_i. Then

$$\tilde{T}^{-1} = (t_{ij}h_i/n),$$

as is easily checked. Now from (41.5) by taking inverses and rearranging

$$I = \psi\tilde{T}^{-1}\phi'. \tag{42.1}$$

Taking the transpose in (41.2′) and multiplying on the left by $\tilde{T}^{-1}$,

$$\tilde{T}^{-1}\psi' = \tilde{T}^{-1}\phi'C' = (\psi)^{-1}C',$$

by (42.1). Thus

$$\psi \tilde{T}^{-1} \psi' = C' = C. \tag{42.2}$$

Again, taking inverses in (41.2′) and (41.5) we get

$$\phi^{-1} C^{-1} = \psi^{-1}, \qquad \psi^{-1} = \tilde{T}^{-1} \phi',$$

which together yield

$$C^{-1} = \phi \tilde{T}^{-1} \phi'. \tag{42.3}$$

Writing out the element at the intersection of the ith row and jth column in (42.1) we have

$$\begin{aligned} \delta_{ij} &= \sum_{\nu=1}^{l} \sum_{\lambda=1}^{l} \psi_i(\breve{g}_\lambda) t_{\lambda\nu} (h_\lambda / n) \phi_j(\breve{g}_\nu) \\ &= (1/n) \sum_{\lambda=1}^{l} \psi_i(\breve{g}_\lambda) h_\lambda \phi_j(\breve{g}_\lambda^{-1}), \qquad \text{since } t_{\lambda\nu} = 0 \text{ if } \lambda \neq \nu \\ &= (1/n) \sum_{g \in G'} \psi_i(g) \phi_j(g^{-1}), \end{aligned}$$

since each character is constant on all h_λ elements of the p-regular class containing $\breve{g}_\lambda$. We rewrite this as

$$(1/n) \sum{}' \psi_i(g) \phi_j(g^{-1}) = \delta_{ij}. \tag{42.1′}$$

Similarly, from (42.2) and (42.3) we obtain

$$(1/n) \sum{}' \psi_i(g) \psi_j(g^{-1}) = c_{ij}, \tag{42.2′}$$

$$(1/n) \sum{}' \phi_i(g) \phi_j(g^{-1}) = \tilde{c}_{ij}. \tag{42.3′}$$

The prime indicates that the summation is only taken over the p-regular elements of G, δ_{ij} is the Kronecker δ, and c_{ij}, $\tilde{c}_{ij}$ are the elements at the intersection of the ith row and jth column of C and C^{-1}, respectively. C is the Cartan matrix.

Example. Let G be the symmetric group $S_3 = \{1, (12), (13), (23), (123), (132)\}$. S_3 can be generated by the elements $a = (12)$ and $b = (132)$. This was our first example at the end of Section 1.

The equivalence classes are $C_1 = \{1\}$, $C_2 = \{(12), (13), (23)\}$, and $C_3 = \{(123), (132)\}$. Thus $s = 3$ and there are three irreducible representations in a field of characteristic zero. We found them to be

$$\sigma_1(a) = \sigma_1(b) = 1; \qquad \sigma_2(a) = -1, \qquad \sigma_2(b) = 1,$$

and

$$\sigma_3(a) = \begin{pmatrix} -1 & 0 \\ -1 & 1 \end{pmatrix}, \qquad \sigma_3(b) = \begin{pmatrix} -1 & 1 \\ -1 & 0 \end{pmatrix}.$$

Let $\Pi = \{\underline{0}, \underline{1}, \underline{2}\}$ be the integers modulo 3. Π is a field of characteristic $p = 3$. Since S_3 has 2 p-regular classes, there are two irreducible modular representations. Clearly,

$$(1) \qquad \begin{cases} \overline{\sigma_1(a)} = \bar{\tau}_1(a) = \underline{1}, & \overline{\sigma_1(b)} = \bar{\tau}_1(b) = \underline{1} \quad \text{and} \\ \overline{\sigma_2(a)} = \bar{\tau}_2(a) = \underline{2}, & \overline{\sigma_2(b)} = \bar{\tau}_2(b) = \underline{1} \end{cases}$$

are irreducible modular representations. Then $\overline{\sigma_3(\;)}$ must be reducible. Now

$$\overline{\sigma_3(a)} = \begin{pmatrix} \underline{2} & \underline{0} \\ \underline{2} & \underline{1} \end{pmatrix}, \qquad \overline{\sigma_3(b)} = \begin{pmatrix} \underline{2} & \underline{1} \\ \underline{2} & \underline{0} \end{pmatrix}, \qquad \text{and if} \qquad T = \begin{pmatrix} \underline{1} & \underline{1} \\ \underline{0} & \underline{2} \end{pmatrix},$$

we find

$$(2) \qquad \begin{cases} T^{-1}\overline{\sigma_3(a)}T = \begin{pmatrix} \underline{1} & \underline{0} \\ \underline{1} & \underline{2} \end{pmatrix} = \begin{pmatrix} \bar{\tau}_1(a) & \underline{0} \\ \underline{1} & \bar{\tau}_2(a) \end{pmatrix}, \\ T^{-1}\overline{\sigma_3(b)}T = \begin{pmatrix} \underline{1} & \underline{0} \\ \underline{1} & \underline{1} \end{pmatrix} = \begin{pmatrix} \bar{\tau}_1(b) & \underline{0} \\ \underline{1} & \bar{\tau}_2(b) \end{pmatrix}. \end{cases}$$

From (1) and (2) it is seen that the decomposition matrix

$$D = \begin{pmatrix} 1 & 0 \\ 0 & 1 \\ 1 & 1 \end{pmatrix}$$

and the Cartan matrix

$$C = D'D = \begin{pmatrix} 2 & 1 \\ 1 & 2 \end{pmatrix}.$$

Note that determinant $C = 3$.

The accompanying tabulation gives the irreducible Brauer characters on the p-regular elements:

	1	a
ϕ_1	1	1
ϕ_2	1	-1

Thus

$$\phi = \begin{pmatrix} 1 & 1 \\ 1 & -1 \end{pmatrix}$$

so that

$$\psi = C\phi = \begin{pmatrix} 3 & 1 \\ 3 & -1 \end{pmatrix}$$

and

$$\chi = D\phi = \begin{pmatrix} 1 & 1 \\ 1 & -1 \\ 2 & 0 \end{pmatrix}.$$

Appendix

1. Groups

(1.1) **Definition.** *A group is a set G together with an operation defined between pairs of elements $a, b \in G$ (the operation is denoted by ab and is called multiplication) which satisfies the following axioms:*

1. *For all $a, b \in G$, $ab \in G$ (closure).*
2. *$(ab)c = a(bc)$ (associativity).*
3. *For each $a, b \in G$ there are elements $x, y \in G$ such that $ax = b$ and $ya = b$ (equations are solvable in G).*

Consequences:

1. Every group has a unique element e with the property $ae = ea = a$, for every $a \in G$ (e is called the identity and is denoted by 1).
2. To every element $a \in G$ there is a unique inverse, written a^{-1}, with the property: $a^{-1}a = aa^{-1} = e$.

If G is a group and if for all $a, b \in G$, $ab = ba$, then G is *commutative* or *abelian*. In this case the operation is usually denoted by $+$, that is, $a + b$ instead of ab. The identity is then written as 0 and the inverse of a as $-a$. A *module* is an abelian group.

(1.2) **Definition.** *A subset S of a group G is called a subgroup if its elements form a group; we must have then: $a, b \in S$ implies $ab \in S$ and $ax = b$, $ya = b$ must be solvable in S for x and y.*

This is equivalent to the conditions: $1 \in S$, and $a \in S$ implies $a^{-1} \in S$.

If S is a subgroup of a group G a left (right) congruence can defined between the elements of G. Thus $a, b \in G$ are left congruent modulo S, written $a \overset{l}{\equiv} b \pmod{S}$ if there is an element $s \in S$ such that $a = sb$. Similarly $a \overset{r}{\equiv} b \pmod{S}$ if $s' \in S$ and $a = bs'$.

It is easy to show that left (right) congruence is an equivalence relation, that is: (1) $a \equiv a$, (2) $a \equiv b$ implies $b \equiv a$, and (3) $a \equiv b, b \equiv c$ implies $a \equiv c$. Then G can be separated into disjoint equivalence classes of left (right) congruent elements. However, we do not in general have that $a \equiv b$ and $c \equiv d$ implies $ac \equiv bd$. For this property to hold S must be a *normal subgroup*

(1.3) **Definition.** *N is a normal subgroup of G, written $N \lhd G$ or $G \rhd N$, if $g^{-1}Ng = N$ for all $g \in G$. That is, if $n \in N$ then $g^{-1}ng = n' \in N$.*

If $N \lhd G$ then left and right congruence modulo N are the same, the elements of G taken modulo N can be multiplied and they form a group called the factor group of G over N, denoted by G/N.

A correspondence between the elements of two groups (or a group with itself) which preserves multiplication is called a homomorphism (endomorphism). More precisely:

(1.4) **Definition.** *A homomorphism of a group G into a group H is a mapping σ which assigns to each element $g \in G$ a unique element $\sigma(g) \in H$, called the image of g, such that for every $g, g' \in G$, $\sigma(gg') = \sigma(g)\sigma(g')$.*

$\sigma(G)$, the set of images of G, is a subgroup of H. If $\sigma(G) = H$, then the homomorphism is said to be *onto*. The homomorphism σ is an *isomorphism* (into or onto in the same way) if distinct elements have distinct images; that is, if $\sigma(a) = \sigma(b)$ if and only if $a = b$. This is equivalent to the condition that $\sigma(a) = 1 \in H$ implies $a = 1 \in G$. We state, without proof, the three Isomorphism Theorems [cf. analogous proofs, Section 2, this appendix].

(1.5) **Theorem I.** *Let σ be a homomorphism of a group G onto a* *group H. Let $N = \{g : \sigma(g) = 1\}$. N is called the kernel of σ. Then $N \lhd G$ and $G/N \cong H$.*

(1.6) **Theorem II.** *If H is a subgroup and N a normal subgroup of a group G, then NH is a subgroup of G, the intersection $H \cap N$ is a normal subgroup of H, and $NH/N \cong H/H \cap N$.* The following diagram is a useful visual aid:

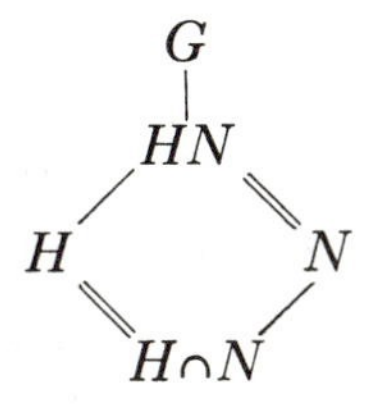

(1.7) **Theorem III.** *Let σ be a homomorphism of the group G onto the group H. Let K be the kernel of σ, so that $K \lhd G$. If $J \lhd H$, and $N = \{g : \sigma(g) \in J\}$, then $K \lhd N \lhd G$ and*

$$G/N \cong H/J \cong G/K\big/N/K.$$

(1.8) **Definition.** *A group G together with a set Ω of endomorphisms (homomorphisms of G into itself) is called a* **group with operators.** *A subgroup S of G is said to be* **admissable**, *if for every $\omega \in \Omega$ $\omega(S) \subset S$.*

Let G and H be two groups with the same domain Ω of operators. A homomorphism (isomorphism) σ of G into H is an *operator homomorphism* (isomorphism) if $\sigma(\omega g) = \omega\sigma(g)$, for every $g \in G$.

Now let G be a group with operators Ω and let

$$G = G_0 \rhd G_1 \cdots \rhd G_i \rhd G_{i-1} \rhd \cdots \rhd G_n = 0$$

be a chain of admissable subgroups of G, each *normal and maximal* in the preceding. Such a chain is called a composition series for G. This series is said to be of *length n*, and the G_i/G_{i-1} are called its *factors*.

A group with a fixed set of operators may have more than one composition series; however, the *Jordan-Hölder theorem* asserts that their lengths must be the same and that the set of factors of any one of them must be isomorphic, in some order, to those of any other.

For references, see: Hall [12], Ledermann [15], Zassenhaus [23], Kurosh [14].

2. Rings, Ideals, and Fields

(2.1) **Definition.** *A set* $R = \{a, b, c, \ldots\}$, *such that for every ordered pair* $a, b \in R$ *there are two operations* $a + b$ *and* ab *called, respectively,* ***addition*** *and* ***multiplication***, *is a ring if:*

(1) *R with respect to addition is an abelian group.*
(2) $a, b \in R$ *implies* $ab \in R$ *and* $(ab)c = a(bc)$, *for every* $a, b, c \in R$.
(3) $a(b + c) = ab + ac$ *and* $(b + c)a = ba + ca$ *(distributive laws).*

If $ab = ba$, for every $a, b \in R$, then R is called a *commutative ring*.

(2.2) **Definition.** *If R is a ring and* $R^* = R - \{0\}$ *is a group with respect to multiplication, then R is a* ***division ring*** *(**s-field, skew-field**) or, in the case that R is commutative, a* ***field***

An element $u \in R$ is a *unit* if it has a multiplicative inverse $u^{-1} \in R$.

(2.3) **Definition.** *If R is a ring, a subset $\mathfrak{A}$ is a right (left) ideal if:*

(1) *$\mathfrak{A}$ is a subring, that is, the elements of $\mathfrak{A}$ themselves form a ring.*
(2) $a \in \mathfrak{A}$ *and* $r \in R$ *implies* $ar \in \mathfrak{A}$ *(*$ra \in \mathfrak{A}$ *if $\mathfrak{A}$ is a left ideal).*

$\mathfrak{A}$ is a two-sided ideal if it is both a left and a right ideal. If R is commutative, every ideal is two-sided.

If A and B are subsets of a ring R we can form their sum $A + B = \{a + b : a \in A, \quad b \in B\}$ and product $AB = \{ab : a \in A, b \in B\}$. If $A = \mathfrak{A}$, $B = \mathfrak{B}$, are left (right, two-sided) ideals of R, then $\mathfrak{A} + \mathfrak{B}$ and $\mathfrak{A}\mathfrak{B}$ are left (right, two-sided) ideals of R. The ideal sum $\mathfrak{A} + \mathfrak{B}$ is usually written $(\mathfrak{A}, \mathfrak{B})$ and is the smallest ideal containing $\mathfrak{A}$ and $\mathfrak{B}$.

Let $\mathfrak{A}$ be a two-sided ideal of R. Define $a \equiv b \pmod{\mathfrak{A}}$, if $a - b \in \mathfrak{A}$. Then $\equiv$ is an equivalence relation on R and separates R into disjoint classes of congruent elements. Let $\lfloor a \rfloor$ denote the class of congruent elements which contain a. The element a is called a representative of this class. Note that $\lfloor a \rfloor = \lfloor a' \rfloor$ if and only if $a \equiv a' \pmod{\mathfrak{A}}$.

It is easy to show that the addition and multiplication of classes defined by $\lfloor a \rfloor + \lfloor b \rfloor = \lfloor a + b \rfloor$ and $\lfloor a \rfloor \lfloor b \rfloor = \lfloor ab \rfloor$ is independent of the representatives chosen to express the class, and that with these operations, the classes form a ring. This is the quotient ring, denoted by $R/\mathfrak{A}$.

(2.4) **Definition.** *Let R and R' be two rings. A mapping σ of R into R' is a **ring homomorphism** if, for every $a, b \in R$*

$$\sigma(a + b) = \sigma(a) + \sigma(b) \qquad \textit{and} \qquad \sigma(ab) = \sigma(a)\sigma(b).$$

*σ is a **ring isomorphism** if distinct elements have distinct images, or equivalently, if $\sigma(a) = 0$ implies $a = 0$.*

In analogy to the isomorphism theorems of groups we have:

(2.5) **Theorem I.** *Let σ be a homomorphism of a ring R onto a ring R'. Let $\mathfrak{A} = \{r \in R : \sigma(r) = 0\}$. Then $\mathfrak{A}$ is a two-sided ideal of R and $R/\mathfrak{A} \cong R'$.*

Proof. Let $a \in \mathfrak{A}$. Then $\sigma(-a) = -\sigma(a) = 0$ and $\sigma(ar) = \sigma(a)\sigma(r) = 0 = \sigma(r)\sigma(a)$. Thus $-a$, ar, $ra \in \mathfrak{A}$. If also $a' \in \mathfrak{A}$ we have $\sigma(a + a') = \sigma(a) + \sigma(a') = 0$ and so $a + a' \in \mathfrak{A}$, which is therefore a two-sided ideal.

Now define $\bar{\sigma}(\lfloor a \rfloor) = \sigma(a)$. Since $\lfloor a \rfloor = \lfloor b \rfloor \Rightarrow a \equiv \pmod{\mathfrak{A}} \Rightarrow a - b \in \mathfrak{A} \Rightarrow \sigma(a - b) = 0 \Rightarrow \sigma(a) = \sigma(b)$, the definition is independent of the representative and is thus unambiguous. Moreover,

$$\begin{aligned} \bar{\sigma}(\lfloor a \rfloor + \lfloor b \rfloor) &= \bar{\sigma}(\lfloor a + b \rfloor) = \sigma(a + b) \\ &= \sigma(a) + \sigma(b) = \bar{\sigma}(\lfloor a \rfloor) + \bar{\sigma}(\lfloor b \rfloor) \end{aligned}$$

and

$$\bar{\sigma}(\lfloor a \rfloor \lfloor b \rfloor) = \bar{\sigma}(\lfloor ab \rfloor) = \sigma(ab) = \sigma(a)\sigma(b) = \bar{\sigma}(\lfloor a \rfloor)\bar{\sigma}(\lfloor b \rfloor)$$

so that $\bar{\sigma}$ is a ring homomorphism of $R/\mathfrak{A}$ onto R'. Finally $\bar{\sigma}(\lfloor a \rfloor) = 0 = \sigma(a) \Rightarrow a \in \mathfrak{A} \Rightarrow \lfloor a \rfloor = \lfloor 0 \rfloor$, the zero element of $R/\mathfrak{A}$ showing that $\bar{\sigma}$ is an isomorphism.

(2.6) **Theorem II.** *If* $\mathfrak{A}$, $\mathfrak{B}$ *are two-sided ideals of* R *then so are* $(\mathfrak{A}, \mathfrak{B})$ *and* $\mathfrak{A} \cap \mathfrak{B}$. *Moreover*, $(\mathfrak{A}, \mathfrak{B})/\mathfrak{A} \simeq \mathfrak{B}/\mathfrak{A} \cap \mathfrak{B}$.

Proof. The first statement is clear. For the second, define $\sigma(\lfloor a + b \rfloor) = \lfloor\lfloor b \rfloor\rfloor$ where the single and double "buckets" refer to classes modulo $\mathfrak{A}$ and $\mathfrak{A} \cap \mathfrak{B}$, respectively. Since $\lfloor a + b \rfloor = \lfloor a \rfloor + \lfloor b \rfloor = \lfloor 0 \rfloor + \lfloor b \rfloor = \lfloor b \rfloor$ this is the same as $\sigma(\lfloor b \rfloor) = \lfloor\lfloor b \rfloor\rfloor$ and is well defined since $\lfloor b \rfloor = \lfloor b' \rfloor \Rightarrow b - b' \in \mathfrak{A} \Rightarrow b - b' \in \mathfrak{A} \cap \mathfrak{B} \Rightarrow \lfloor\lfloor b \rfloor\rfloor = \lfloor\lfloor b' \rfloor\rfloor$. Now, as before, we can show that σ is a ring homomorphism of $(\mathfrak{A}, \mathfrak{B})$ onto $\mathfrak{B}/\mathfrak{A} \cap \mathfrak{B}$ and then Theorem I of 2.5 gives the result.

For references, see: Artin [2]; van der Waerden [22].

3. A Formula for the Character

The character of the representation corresponding to the idempotent derived from PN (Lemma 28.6) can be calculated by the formula

$$\text{(3.1)} \qquad \chi(g) = (n/(\mathfrak{C}(g) : 1)(R \cap C : 1)) \sum_{rc \in \mathfrak{C}(g)} \theta(r)\phi(c)$$

where χ is the character of the irreducible representation corresponding to the primitive idempotent formed from R and C, n is the degree of the irreducible representation, $\mathfrak{C}(g)$ is the class of elements conjugate to g, and the other symbols are as in Lemma 28.6.

Proof. It is clear that $\sum_{s \in G} s(PN)s^{-1}$ is an element of the center of the subalgebra to which PN belongs; moreover the expression

$\sum_{t \in G} t\chi(t)$ is the central idempotent of this subalgebra up to a multiple. Thus, since PN remains a multiple of a primitive idempotent even after extension of the ground field to an algebraically closed field so that the center of the Wedderburn component is of dimension 1,

$$f \sum_{t \in G} t\chi(t) = \sum_{s \in G} s(PN)s^{-1}$$

Recalling that $PN = \sum rc\theta(r)\phi(c)$ and equating coefficients of g on both sides we get $f\chi(g) = \sum' \theta(r)\phi(c)$, where the summation is over all r, c for which, for some s, $srcs^{-1} = g$. Now if the relation holds for a particular element s, then it holds also for the element hs if h is an element of the normalizer $\mathfrak{N}(g)$ of g: $hs(rc)(hs)^{-1} = hgh^{-1} = g$. It follows that the contribution to the sum for each r, c for which $rc \in \mathfrak{C}(g)$ is repeated $\mathfrak{N}(g) : 1$ times. This permits us to write

$$f\chi(g) = (\mathfrak{N}(g) : 1) \sum_{rc \in \mathfrak{C}(g)} \theta(r)\phi(c) \tag{3.2}$$

In particular if g is the identity element I of the group G then $\mathfrak{N}(g) : 1 = G : 1$ and $rc = I$ so that r and c must be from $R \cap C$ and the condition of Lemma 28.6 requires that $\theta(r) = \phi^{-1}(c)$. Consequently $fn = (G : 1)(R \cap C : 1)$ where $n = \chi(I)$ is the degree of the irreducible representation. Substitution for f in (3.2) gives the result (3.1) if we recall that $(\mathfrak{C}(g) : 1)(\mathfrak{N}(g) : 1) = G:1$.

Bibliography

List of titles referred to in the text. For an extensive Bibliography, see [10].

1. Artin, E., Nesbitt, C., and Thrall, R. M., "Rings with Minimum Condition." Univ. of Michigan, Ann Arbor, Michigan, 1944.
2. Artin, E., Theory of Algebraic Numbers, Lecture notes, Goetingen, 1956.
3. Brauer, R., and Nesbitt, C., "On the Modular Representations of Finite Groups." Univ. of Toronto Studies, Math. Ser. Vol. 4 (1937).
4. Brauer, R., Sur Darstellungstheorie der Gruppen endlicher Ordnung, *Math. Z.* **63** (1956), 406–444.
5. Brauer, R., A characterization of the characters of groups of finite order, *Ann. Math.* **57** (1953), 357–377.
6. Brauer, R., and Tate, J., On the characters of finite groups, *Ann. Math.* **62** (1955), 1–7.
7. Burnside, W., "Theory of Groups of Finite Order," 2nd ed. Cambridge Univ. Press, Cambridge, 1911.
8. Burrow, M. D., A generalization of the Young diagram, *Canad. J. Math.* **6** (1954), 498–508.
9. Coxeter, H. S. M., and Moser, W. O. J., "Generators and Relations for Discrete Groups." Springer, Berlin, 1957.
10. Curtis, W. C., and Reiner, I., "Representation Theory of Finite Groups and Associative Algebras." Wiley (Interscience), New York, 1962.
11. Dickson, L., "Linear Groups." Dover, New York.
12. Hall, M., "The Theory of Groups." Macmillan, New York, 1959.
13. Higman, D. G., Indecomposable representations at characteristic p, *Duke Math. J.* **21** (1954), 377–381.
14. Kurosh, A., "The Theory of Groups," 2 vols. Chelsea, New York, 1955.
15. Ledermann, W., "The Theory of Finite Groups." Oliver & Boyd, Edinburgh and London, 1953.
16. MacDuffee, C. C., "An Introduction to Abstract Algebra." Wiley, New York, 1948.
17. Osima, M., Note on blocks of group characters, *Math. J. Okayama Univ.* **4** (1955), 175–188.
18. Ree, R., On generalized conjugate classes in a finite group, *Illinois J. Math.* **3** (1959), 440–444.

19. Robinson, G. de B., "Representation Theory of the Symmetric group." Univ. of Toronto Press, Toronto, 1961.
20. Rutherford, D. E., "Substitutional Analysis." Oliver & Boyd, Edinburgh and London, 1948.
21. Speiser, A., "Die Theorie der Gruppen von endlicher Ordnung." 3rd ed. Springer, Berlin, 1937.
22. van der Waerden, B., "Modern Algebra." Ungar, New York, 1949.
23. Zassenhaus, H., "The Theory of Groups." Chelsea, New York, 1958.
24. Zassenhaus, H., Trace functions on algebras with prime characteristic, *Amer. Math. Monthly* **60**, 10 (1953), 685–692.
25. Feit, W., and Thompson, J. G., Solvability of Groups of odd order. *Pacific J. Math.* **13**, 3 (1963), 775–1029.

Subject Index